NAKED EGGS
<small>AND</small> FLYING POTATOES

UNFORGETTABLE EXPERIMENTS
THAT MAKE SCIENCE FUN

STEVE SPANGLER

GREENLEAF
BOOK GROUP PRESS

By their very nature, science experiments fizz, bubble, pop, smoke, erupt, move, change temperature, and sometimes produce unexpected results. That's why science is fun, and that's why you need to follow the necessary safety precautions when doing any science activity.

   The material in this book is provided for informational, educational, noncommercial, and personal purposes only and does not necessarily constitute a recommendation or endorsement of any company or product.

   While the science experiments and demonstrations in this book are generally considered safe and a low hazard, please use care when performing any science experiment. Adult supervision of kids is always recommended. If you're an adult who acts like a kid, find a more responsible adult to supervise you. Common sense and care are essential to the conduct of any and all activities, whether described in this book or otherwise. Without limitation, no one should ever look directly at the sun, attempt to build a time machine, or eat yellow snow.

   The author and publisher expressly disclaim all liability for any occurrence, including, but not limited to, damage, injury, or death, which might arise from the use or misuse of any project or experiment in this book. Wow . . . this is serious.

Published by Greenleaf Book Group Press
Austin, Texas
www.gbgpress.com
Copyright ©2010 Steve Spangler

Distributed by Greenleaf Book Group LLC

For ordering information or special discounts for bulk purchases, please contact
Greenleaf Book Group LLC at PO Box 91869, Austin, TX 78709, 512.891.6100.

Design and composition by Greenleaf Book Group LLC
Cover design by Greenleaf Book Group LLC
Edited by Debbie Leibold
Photography by Shawn Campbell and Bradley Mayhew

Publisher's Cataloging-In-Publication Data
(Prepared by The Donohue Group, Inc.)
Spangler, Steve.
   Naked eggs and flying potatoes : unforgettable experiments that make science fun / Steve Spangler. -- 1st ed.
   p. : col. ill. ; cm.
   ISBN: 978-1-60832-060-8

   1. Scientific recreations--Handbooks, manuals, etc. 2. Science--Experiments--Handbooks, manuals, etc. I. Title.
Q164 .S62 2010
507.8   2010928622

Part of the Tree Neutral™ program, which offsets the number
of trees consumed in the production and printing of this book
by taking proactive steps, such as planting trees in direct
proportion to the number of trees used: www.treeneutral.com

TreeNeutral™

Manufactured by Oceanic on acid-free paper
Manufactured in Guang Dong Province, China, February 2015
Batch No. TT14120651-R01

15 16 17 18   10 9 8 7 6 5

First Edition

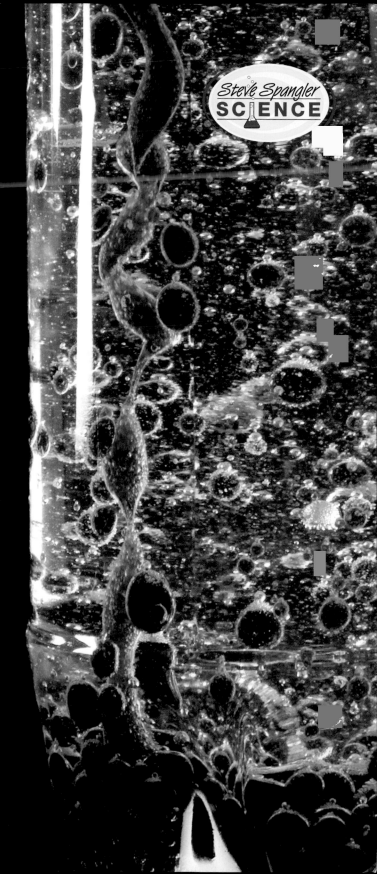

CONTENTS

DRY ICE   93

GOOEY WONDERS   103

DON'T TRY THIS AT HOME . . .
TRY IT AT A FRIEND'S HOME!   123

THANKS FOR CREATING A FEW
UNFORGETTABLE LEARNING EXPERIENCES   155

IT'S NOT ABOUT THE SCIENCE ... **IT'S ABOUT THE EXPERIENCE**

I've worked in television for many years, but not as the evening anchor or sports guy or even the weatherman. When the little red light comes on, it's my job to teach viewers how to do amazing things using ordinary stuff found around the house. What type of amazing things, you ask? Oh, things like how to make a high-powered potato launcher out of pencils and straws or how to make a 2-liter bottle of soda erupt into a 12-foot-high fountain of fun. That's right, you guessed it—when the red light comes on, I become the science guy—a modern-day "Mr. Wizard," so to speak.

My first job, however, was not in television. Fresh out of college with a teaching certificate in hand, I found a job in an elementary school teaching science. It didn't take

long for me to discover that my style of teaching was somewhat different from that of my colleagues, who spent most of their time running off worksheets in the copy room. My kids laughed a lot (almost too much at times), and this soon caught the attention of neighboring teachers and their kids who were more than a little curious. One colleague asked, "How can your students be learning when they're laughing so much?" Hmmm . . . I wonder if laughter and learning go hand in hand? The answer is yes!

I have to attribute most of my success as a teacher to my first class of third graders. Over the course of 9 months, they taught me the importance of using humor to create experiences that transcend the four walls of the classroom and somehow make it to the dinner table as a topic of conversation.

"What did you do in school today?"

"Not much. Oh . . . I remember something . . . Mr. Spangler made us get into a big circle and hold hands . . . then he shocked us with 50,000 volts to teach us about electricity."

> "OH . . . I REMEMBER SOMETHING . . . MR. SPANGLER MADE US GET INTO A BIG CIRCLE AND HOLD HANDS . . . **THEN HE SHOCKED US WITH 50,000 VOLTS** TO TEACH US ABOUT ELECTRICITY."

I got lots of calls from parents that first year of teaching, and it didn't take long for word to spread that things were a little different in the new teacher's class. One of those parents just happened to work for the local NBC television affiliate in Denver. She invited me to come down to the station some day after school and asked if I would bring along a few science experiments from my class, including that shocking machine. In no time, I had a group of television executives making slime, shooting potatoes, and holding hands in a big circle while I delivered the shock. That one commanding performance opened the door for me into a much bigger classroom. I went from twenty-three kids to over a million viewers each week as the host of a nationally syndicated children's program called *News for Kids*.

My executive producer spelled out my mission in the clearest terms possible: "Your job is not to teach science. Your job is to grab the viewer's attention and show them that learning is fun. Make them laugh and the learning will follow." These marching orders soon became my mantra and the advice that I give to parents and teachers today.

3-2-1 BLAST OFF!

As part of a promotional tour for the television show, I found myself on the road, visiting children in schools across the country with my bag of cool gadgets and science demonstrations. Let's just say that there's nothing terribly glamorous about doing school assemblies. The best-case scenario is that a bunch of kids are crammed into the cafeteria and forced to sit on the floor, while the guest speaker is forced to shout because the P.E. teacher is using the microphone as a doorstop. On this particular occasion, the setting was an elementary school in the heart of Salt Lake City. Nearly seven hundred children squeezed their way into the cafeteria, and the principal's introduction was nothing short of inspirational.

"Hey kids . . . listen up. There's a guy here who wants to show you something and I want you to be good for a change. If I catch anyone throwing stuff at the speaker like you did last time, I'm shutting this circus down." He turned to me. "Okay . . . they're all yours."

With an introduction like that, things could only get better. Up to this point, I had never really taught kindergartners, but I soon learned that these little people have a tendency to grab onto you as a sign of affection! I did most of the show with a five-year-old latched onto my leg. Thankfully, the kids liked the demos and I survived my first of two presentations. As the sea of children started to file out of the room, I noticed that one of the kindergartners was not ready to leave. In fact, he wanted to talk to me. As he approached, I could tell that he was a little nervous. He pulled at his pant leg and squirmed as if it might be time to find a bathroom. As I kneeled down, he began to talk.

"Ummm . . . hey guy. Guess what?"

"What?"

"I like rockets."

"Me, too!"

"And you know what else?" he said. "I know how to make a rocket . . . and someday I will make a rocket that can fly to the sun!"

THEN THE LIGHTBULB IN MY HEAD WENT ON. BEHIND EVERY FUNNY KINDERGARTNER THERE'S A FUNNIER PERSON CALLED A TEACHER.

Well, here's a tough fork in the road. I can't tell him "no" because I would crush his dream, and I can't say "great" because I would be lying. They just don't teach you this stuff in college. I looked him right in the eyes, and with compassion in my voice I said, "I like your idea, but if your rocket gets too close to the sun, it will melt."

He looked at me the way only a kindergartner could and said, "I'm doing it at night, duh!" It was as if I had swallowed the bait and he was reeling in the catch of the day. The best part is that I had heard someone tell me the same joke years before, but I had never heard it told by a kindergartner!

Then the lightbulb in my head went on. Behind every funny kindergartner there's a funnier person called a teacher. I immediately looked over the sea of kids to find his kindergarten teacher looking right at me with a huge grin as she

mouthed the phrase, "Gotcha!" I turned my attention back to the little comedian and said, "You are so funny!" His reply was phrased with a sense of apprehension, "I don't know why everyone thinks that joke is so funny."

What? Didn't the kid get it? Then it hit me like a ton of bricks. This little boy still *believed*. In his way of thinking, *all things are possible*. What was so funny to me and to his teacher offered little in the way of humor to him because his world was filled with limitless possibilities.

Maybe this belief is my real reason for writing this book. You don't have to understand that a rocket can't travel to the sun. You don't have to know what osmosis is or be able to quote Newton's First Law of Motion. The science behind the activities doesn't really matter (okay, the science is important), what truly matters is the experience.

To the casual observer, this book represents a collection of cool science activities that might be considered *geek chic*. There's just something cool about learning how to remove an egg's shell to reveal a naked egg, or how to use a bar of soap to whip up a soap soufflé in your friend's microwave oven.

> **THE FIRST TIME YOU TRY ONE OF THE ACTIVITIES YOU'LL DISCOVER THAT THERE'S AN INFECTIOUS, ALMOST VIRAL QUALITY THAT TRANSFORMS THE "COOL ACTIVITY" INTO AN UNFORGETTABLE LEARNING EXPERIENCE.**

The first time you try one of the activities, however, you'll discover that there's an infectious, almost viral quality that transforms the "cool activity" into an unforgettable learning experience. When you read about the science behind making a bottle of soda erupt into a geyser using Mentos, you'll be compelled to try it yourself. But it doesn't stop there. The next thing you know, you're sharing what you just experienced with a friend . . . and then it hits you . . . you've caught the bug.

What is the bug? It is a renewed sense of wonder. You can't stop asking questions, wondering "What would happen if . . . ?" and trying new variations on the activities. Your enthusiasm is contagious and your friends "catch it" too.

This is my hope as you read this book. Catch the bug. Wonder. Ask questions. Share your discoveries. Make these activities unforgettable learning experiences. And more than anything else, believe that the possibilities are endless.

—Steve

A WORD ABOUT **SAFETY**

By their very nature, science experiments fizz, bubble, pop, smoke, erupt, move, change temperature, and sometimes produce unexpected results. That's why science is fun, and that's also why you need to follow the necessary safety precautions when doing any science activity.

Read all the directions before you begin any experiment, and if you aren't sure about something, ask someone who knows!

Ask an adult to help you when you use sharp utensils, heat sources, or chemicals. Take it seriously when the experiments say that they require adult supervision.

Don't put any chemical near your mouth, eyes, ears, or nose. The incorrect use of chemicals can cause injury and damage your health.

Wear safety glasses when the instructions tell you they're necessary.

Keep the area surrounding your activity clear of any obstructions and keep the contents of the activity away from food or food storage areas. Your workspace should be well-lit, ventilated, and close to water.

Wear protective gloves or use tongs when handling dry ice because it will cause severe burns if it comes in contact with your skin. Never put dry ice into your mouth! Never trap dry ice in a container without a vent.

Wash your hands thoroughly with soap and water after handling raw eggs. Some raw eggs contain salmonella bacteria that can make you really sick.

Do not pour polymer mixtures (goo and slime) down the drain. Dispose of them in the garbage can.

Watch out for small pieces, balloons, or rubber bands that may be part of your experiment. Not only are they easy to lose, they can also be a choking danger to young children or pets who might try to swallow them.

Aim anything that is going to "shoot" or explode away from yourself, other people (even your little brother), animals, and breakable items.

Use the ingredients only for what the instructions describe.

Adults should exercise discretion as to which experiments are suitable and safe for their children.

Don't eat your science experiments . . . they don't taste good and it's a bad thing to do.

Bottom line—the science activities in this book require adult supervision (some adults even need adult supervision) and common sense, simple essentials that will help ensure a fun and safe experience.

THE POWER
OF **AIR**

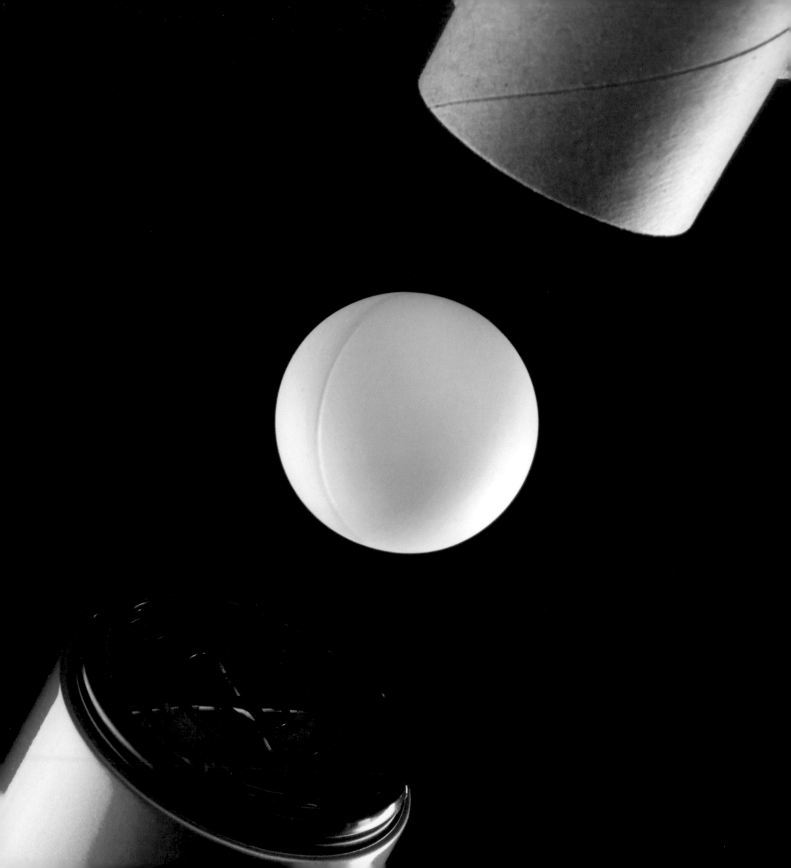

FLOATING PING-PONG BALLS AND FLYING TOILET PAPER

Amuse the neighbors for hours as you make objects float in midair. Believe it or not, the secret to this levitation mystery is right in front of your nose.

LEGAL NOTICE

The reader hereby agrees to only use the powers of flying toilet paper for good, not evil, purposes. The reader also promises never to TP (Toilet Paper) a friend's house using the Flying Toilet Paper Method described herein. However, if an accidental "TP-ing" occurs, please send the author pictures from your cell phone and a full account of what happened.

LET'S TRY IT!

1. Set the hair dryer to cool, switch it on, and point it at the ceiling.

2. Carefully put the ping-pong ball in the stream of air. Hold the hair dryer very steady and watch as the ping-pong ball floats in the stream of air.

3. Carefully move the hair dryer from left to right and watch how the ball moves as well, staying in the stream of air.

4. Try floating other lightweight objects in the air stream at the same time. With the hair dryer on, place an inflated balloon over your levitating ping-pong ball. You might want to place a penny in the balloon before you blow it up to give it some added weight.

TAKE IT FURTHER

Try to float two or more balls in the same air stream. How many can you float at once? How do they behave when there is more than one?

Need more power? Try using a leaf blower in place of the hair dryer. Now you can float larger objects like beach balls.

Want to make flying toilet paper? Just hold a roll of toilet paper in the stream of air and watch the paper take off! Be sure to hold the toilet paper roll on a long stick (piece of dowel) in order for it to spin fast and unroll the paper.

And for the finale . . . balance a ping-pong ball in the air stream.

Then place your now empty toilet paper tube above it in the air. Watch it float above the ball. Then watch the ball get sucked up inside the toilet paper tube. TA DA!! Always conclude this demo with thanks to Bernoulli (see below if you don't get it).

WHAT'S GOING ON HERE?

The floating ping-pong ball is a wonderful example of **Bernoulli's Principle**. Bernoulli, an 18th-century Swiss mathematician, discovered something quite unusual about moving air. He found that the faster air flows over the surface of something, the less the air pushes on that surface (and so the lower its pressure). The air from the hair dryer flows around the outside of the ball and, if you position the ball carefully, the air flows evenly around each side. Gravity pulls the ball downward while the pressure below the ball from the moving air forces it upward. This means that all the forces acting on the ball are balanced and the ball hovers in midair.

As you move the hair dryer you can make the ball follow the stream of air because Bernoulli's Principle says that the fast moving air around the sides of the ball is at a lower pressure than the surrounding stationary air. If the ball tries to leave the stream of air, the still, higher pressure air will push it back in. So, the ball will float in the flow no matter how you move.

When you place the empty toilet paper tube into the air stream, the air is funneled into a smaller area, making air move even faster. The pressure in the tube becomes even lower than that of the air surrounding the ball, and the ball is pushed up into the tube.

REAL-WORLD APPLICATION

Airplanes can fly because of Bernoulli's Principle. Air rushing over the top of airplane wings exerts less pressure than air from under the wings. So the relatively greater air pressure beneath the wings supplies the upward force, or lift, that enables airplanes to fly.

STRAW THROUGH **POTATO**

Sometimes you have to stop and ask yourself, "Who comes up with this stuff?" No one ever uses a straw to eat a potato, but science aficionados seem to enjoy finding ways to poke straws through potatoes anyway. There must be a deeper meaning . .

WHAT YOU NEED

Two or more stiff straws

A big, raw potato

Paper towels to clean up afterward

LET'S TRY IT!

1. The challenge is quite simple: Stab the straw through the potato without bending or breaking the straw. Most of your guests will think it can't be done, but you, of course, know better.

2. As you hold the potato, keep your fingers on the front and your thumb on the back and not on the top and bottom. You don't want to stab yourself! Grab the straw with your writing hand and (shhh! this is the secret) cap the top end with your thumb.

3. Hold on firmly to both the straw and the potato and with a quick, sharp stab, drive the straw into and partway out of the narrow end of the spud (not the fatter, middle part).

4. Your audience will be impressed and will want to try it. Tell them to hold the spud the way you did so they don't stab a finger or thumb with the straw. They may not know the secret, but don't give it away just yet. Let them try to figure it out. You may need more stiff straws.

TAKE IT FURTHER

You know exactly where to go next. Open the refrigerator and search for your favorite fruits and vegetables that seem to be best suited for a straw attack.

WHAT'S GOING ON HERE?

The real secret is inside the straw—it's air. Placing your thumb over the end of the straw traps the air inside. When you trap the air inside the straw, the air molecules compress and give the straw strength, which in turn keeps the sides from bending as you jam the straw through the potato. The trapped, compressed air makes the straw strong enough to cut through the skin, pass through the potato, and exit out the other side. Without your thumb covering the hole, the air is simply pushed out of the straw and the straw crumples and breaks as it hits the hard potato surface.

Be sure to keep your fingers out of the way. After you stab with the straw, take a look at the end that passed through the potato. There's a plug-o'-spud inside the straw. If you should have a finger or thumb or hand in the way of the straw as it collides with the potato, then there will be a plug-o'-you in the straw too. Ouch!

FLYING **POTATOES**

What could be more fun than a potato gun, a 14-pound sack of spuds, and a big cup of coffee? Okay, it's not really a "gun"—more of a potato pusher. But be careful . . . in the midst of your fun, you might learn a bit of science too. What a concept!

LET'S TRY IT!

1. There are two parts to the potato launcher—the plunger (your dowel rod) and the PVC tube. Let's start with the plunger. Form a stopper by wrapping a 12-inch strip of duct tape around the dowel rod approximately 6 inches from the end of the rod. Do not wrap any tape around the second dowel rod. You'll use this rod as a way to push any remaining pieces of potato out of the PVC tube when you're finished with the activity.

2. Now let's move on to the PVC tube. Most versions of the potato launcher use opaque PVC tubing like the kind you would find at the hardware store. However, clear PVC tubing is available if you want to see what is happening inside the tube. Clear PVC is much more expensive than traditional opaque PVC, but it looks great. In order for a piece of potato to fit tightly in the tube, it's necessary to flare *both* ends of the PVC pipe. One of the easiest methods for doing this involves an electric drill and a cone-shaping grinding stone, but if you don't have these tools, use a knife to flare the edges with the help of an adult. Use care as the ends of the tube can be sharp or have rough edges.

3. Let's not forget about the potato. Place the potato on a flat surface. If the potato is very thick, you might want to cut it in half lengthwise. Hold the potato securely with one hand while pushing the end of the tube through the potato with the other hand. Pull the tube out of the potato

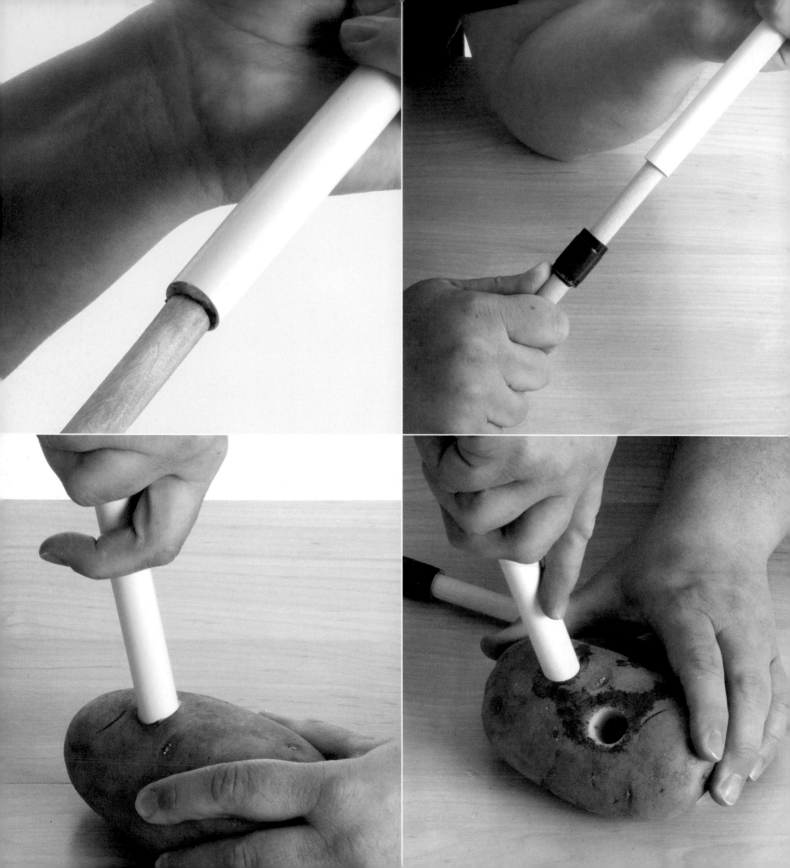

to see your "potato plug." (The sharpened, flared edges of the tube should help cut through the potato.)

4. Hold the plunger (dowel rod) with your hand behind the stopper you made out of duct tape. Use the plunger to move the piece of potato to the other end of the tube (the tape stopper will actually position the potato plug a few inches from the other end). The photograph of the clear PVC tubing shows the proper positioning of the potato plug. Always keep your pushing hand away from the sharp edge of the tube and positioned behind the tape stopper. Remove the plunger.

5. The reason for moving the potato plug is to free up the end of the tube to accept another piece of potato. Position the potato securely on a flat surface while pushing the flared edges of the empty end of the tube into the potato again. Now both ends of the potato tube are plugged.

6. Assuming that you're right handed, hold the plastic tube in the middle with your left hand and the plunger in your right hand. The plunger will go into the end where the first potato plug is a few inches from the end of the tube.

7. Push upward with the plunger on the bottom piece of potato until the top potato piece pops out of the tube. POW! Because of the placement of the duct tape stopper, the bottom potato should now be positioned a few inches from the top of the tube, and the bottom end of the tube is ready to accept another unsuspecting piece of potato.

8. It takes the average potato-launching science enthusiast about thirty launches before feeling completely confident about the mission. Never aim the flying potato at anyone. It's best to do this demo outside, away from all forms of life. Remember, this is a science demonstration, not a bombing mission. Use caution when demonstrating your newly acquired skill.

9. When you are finished, use the second dowel rod without the duct tape stopper to carefully push the remaining potato plug out of the tube. Wash and rinse the tube and the plunger with mild soap and water.

WHAT'S GOING ON HERE?

The potato gun beautifully illustrates **Boyle's Law,** which states that pressure and volume are inversely proportional. In other words, as you decrease the volume of the air trapped in between the two pieces of potato, the pressure exerted by the air increases. This increase in pressure eventually forces the potato at the top end to exit the tube with great pizzazz!

DO NOT OPEN BOTTLE

When you receive a package that says "DO NOT OPEN!" what do you want to do? Open it!
It's just human nature. Try this tempting experiment for some guaranteed fun.

WHAT YOU NEED

Clear plastic soda bottle
(1 liter with cap)

Permanent marker, any color

Sharp pushpin

Water

Flashlight

Towel to clean up your mess

LET'S TRY IT!

1. Clean and dry the 1-liter bottle and remove the label.

2. Fill the bottle to the very top with water and twist on the cap.

3. Use the permanent marker to write "DO NOT OPEN!" in fat letters on the bottom half of the bottle.

4. Carefully, use a sharp pushpin to poke a line of five or six holes about an inch (2.5 cm) from the bottom of the bottle. A small amount of water will squirt out as you poke holes in the bottle, but it's not a big deal. When you're finished, hold the bottle by the cap (don't squeeze the bottle or it will start leaking before you're ready) and give the bottle a gentle wipe down with the towel.

5. Carefully set the bottle on the kitchen counter (word-side out) where someone can see it as they pass by. Stay close enough to watch what happens. Eventually, someone is bound to ask about the bottle. Play dumb with, "I dunno," when they ask about it. Let them unscrew the cap and you'll witness science in action. Water squirts everywhere!

6. As you test out your newly discovered water management skills, you'll quickly notice that it's funnier to watch people just pick up the bottle. Even the slightest squeeze on the sides as they lift the bottle results in water squirting from the holes. Yes, it's childish . . . but it's really fun . . . and it's a great science lesson.

TAKE IT FURTHER

Experiment by poking different numbers of holes in the bottle. Do you get the same result with twenty holes as you did with just five holes? Try poking holes in different areas of the bottle. Do holes in the top of the bottle leak the same amount of water as holes in the bottom of the bottle? Does the size of the hole matter?

Leaking Liquid Light

There's a good use for your DO NOT OPEN bottle. Yes, it's a great practical joke, but there's also some real science waiting to pour out.

Place the bottle on the counter close to the sink so the streams of water flow into the sink. You'll need to get a big, powerful flashlight to shine through the bottle of water. Turn on the flashlight and turn off the lights in the room. Now, uncap the lid and let the water begin to flow.

As the light from the flashlight passes through the bottle, the water acts like a pipe for the light. Place your hand under the stream of water and look for the spot of light on the palm of your hand. Amazing!

The light is actually trapped inside the flowing stream of water. The water also helps to reflect the beam of light back into the stream whenever it tries to escape. When the light stream shoots down the stream of water, the waves of light bounce off the sides of the stream, focusing it back toward the middle. This idea of light bouncing around the inside of the stream is called **total internal reflection**.

Why is this important? You might have heard of your telephone company stringing fiber optic cables in place of the traditional metal core wires. Fiber optic cables are made out of glass fibers. A laser beam that carries your telephone conversation is fired into one end of the cable where it bounces off the internal walls and comes out the other end at about the speed of light. Your soda bottle light pipe is a great way to start to understand the physics of fiber optics. If you want to learn more about fiber optics, make a trip to your local library or start your scientific surfing on the Internet.

WHAT'S GOING ON HERE?

Let's start by examining an empty soda bottle. Is the bottle really empty? No. The bottle is filled with air (gotcha!). When you pour water into the bottle, the molecules of air that once occupied the bottle come rushing out of the top. You don't notice this because molecules of air are invisible. When you turn a bottle filled with water upside down, the water pours out (thanks to gravity) and air rushes into the bottle. Think of it as an even exchange of water for air.

You might think that poking a tiny hole in the bottom of a bottle would cause it to leak, and it does if air molecules can sneak into the bottle. When the lid is on the soda bottle, air pressure can't get into the bottle to push on the surface of the water. The tiny holes in the bottom or sides of the bottle are not big enough for the air to sneak in. Believe it or not, the water molecules work together to form a kind of skin to seal the holes—it's called **surface tension**. When the lid is uncapped, air sneaks in through the top of the bottle and pushes down on the water (along with the force of gravity), and the water squirts through the holes in the bottle. So be patient, observe, and wait for the scream. Be prepared to say, "Can't you read? It says, 'Do Not Open!'" Then run to get the towels.

THE INCREDIBLE CAN CRUSHER

There are lots of different ways to crush a soda can—with your foot, in your hands, on your head. But nothing compares to the fun you'll have doing the soda can implosion experiment. Just wait until the can goes "POP!" Then you'll see who has nerves of steel.

LET'S TRY IT!

1. Start by rinsing out the soda cans to remove any leftover soda goo.

2. Fill the bowl with cold water (the colder the better).

3. Add 1 generous tablespoon of water to the empty soda can (just enough to cover the bottom of the can).

4. Place the can directly on the burner of the stove while it is in the "OFF" position. It's time for that adult to turn on the burner to heat the water. Soon you'll hear the bubbling sound of the water boiling and you'll see the water vapor rising out from the can. Continue heating the can for one more minute.

5. It's important to think through this next part before you do it. Here's what's going to happen: you're going to use the tongs to lift the can off the burner, turn it upside down, and plunge the mouth of the can down into the bowl of water. Get a good grip on the can near its bottom with the tongs, and hold the tongs so that your hand is in the palm up position. Using one swift motion, lift the can off the burner, turn it upside down, and plunge it into the cold water. Don't hesitate . . . just do it!

6. Wow, and you thought that you had nerves of steel. The can literally imploded. Before you jump ahead to the explanation, stop to ponder how this works. What force is great enough to crush the can?

7. Don't just sit there . . . get back to that stove and do it again! Each time you repeat the experiment, carefully observe what is happening in order to try to figure out what's going on.

WHAT'S GOING ON HERE?

Here's the real scoop on the science of the imploding can. Before heating, the can is filled with water and air. By boiling the water, the water changes states from a liquid to a gas. This gas is called water vapor. The water vapor pushes the air that was originally inside the can out into the atmosphere. When the can is turned upside down and placed in the water, the mouth of the can forms an airtight seal

against the surface of the water in the bowl. In just a split second, all of the water vapor that pushed the air out of the can and filled up the inside of the can turns into only a drop or two of liquid, which takes up much less space. This small amount of condensed water cannot exert much pressure on the inside walls of the can, and none of the outside air can get back into the can. The result is the pressure of the air pushing from the outside of the can is great enough to crush it.

The sudden collapsing of an object toward its center is called an **implosion**. Nature wants things to be in a state of equilibrium or balance. To make the internal pressure of the can balance with the external pressure on the can, the can implodes. That's right, air pressure is powerful!

One more thing . . . you probably noticed that the can was filled with water after it imploded. This is a great illustration of how air is pushing all around us. Specifically,

the outside air pressure was pushing downward on the surface of the water. Since the air pressure inside the can was less than the pressure outside the can, water from the bowl was literally pushed up and into the can.

This action is similar to what happens when you drink from a straw. Though we say we are "sucking" liquid up through the straw, we really aren't. To put it simply, science doesn't suck . . . it just pushes and pulls. Outside air pressure is pushing down on the surface of the liquid. When you reduce the pressure in your mouth (that sucking action) the outside pressure is greater than the pressure inside your mouth and the soda shoots up through the straw and into your mouth. The same thing is true with the can. The outside air pressure pushing downward on the surface of the water is greater than the force inside the can and the water gets pushed up into the can.

Crushing soda cans is fun, but I invited some people to help me take this experiment to a whole new level. In place of a soda can, we used a steel 55-gallon drum, a propane camping stove, and a kid's swimming pool filled with ice water.

HOW TO MAKE
GIANT SMOKE RINGS

Years ago, toy manufacturers sold air blasters that sent bursts of air sailing across a room to the surprise and delight of any innocent victim. With a little practice, it was quite easy to shoot a cup off of someone's head from 20 feet away. I took this idea to the next level using the motto, "Make It Big, Do It Right, Give It Class!" and created the Trash Can Smoke Ring Launcher, which turned into one of my signature demonstrations. There's both a small and a large version of the demonstration depending on just how much fun you want to have. Here's my advice . . . go for the trash can version!

WHAT YOU NEED

5-gallon bucket or a
large trash can

Bungee cord

Plastic shower curtain or
thick plastic sheet

Knife, keyhole saw, or
drill with a hole saw

LET'S TRY IT!

1. The smaller version of the air blaster uses a 5-gallon plastic bucket. Carefully cut a 2- to 3-inch hole in the bottom center of the bucket. Use care when cutting the hole with a knife or keyhole saw.

2. Stretch a membrane across the top of the bucket. A piece of clear plastic shower curtain works great. Just stretch a piece of the shower curtain or plastic sheet over the top of the bucket and secure it in place using a bungee-type cord.

3. Lightly hit the shower curtain with your hand or the end of a stick. An invisible blast of air shoots out of the hole. Just aim the air cannon at someone or something across the room and send a blast of air their way with a whack of the membrane.

4. The larger version uses a plastic trash can in place of the smaller bucket. Carefully cut an 8-inch diameter hole in the bottom of the trash can. Stretch a piece of clear shower curtain or a thick plastic sheet over the top of the trash can and secure it in place using a large bungee-type cord. Aim the hole in the bottom of the trash can at your victim and smack the membrane. The blast of air is strong enough to really startle the unsuspecting person.

TAKE IT FURTHER

How can you make the invisible blast of air visible? Try adding a little smoke. Believe it or not, the so-called blast of air is actually shaped like a ring, and just a little smoke will make the rings visible.

How to Make a Smoke Ring Launcher

Smoke rings are made by filling the bucket or trash can with a little theatrical smoke. Just position the hole in the trash can or bucket up against the smoke

machine and give it a blast. Smoke machines (foggers) are commonly used in stage productions and are available at department stores at Halloween. You just need enough smoke to fill the container—a quick blast will do the trick.

Aim the hole in the bottom of the container up into the air and *gently* tap the shower curtain (a hard smack results in a fast blast of air that is difficult to see). The flying vortices are best seen against a dark background with light coming from either side. With a little practice, you can use the power of the smoke ring to knock a cup off of someone's head at a distance of 20 to 30 feet!

Fourth of July Smoke Bomb Twist

The smoke machine is the traditional way of creating smoke rings in your air blaster, but there's no need for a smoke machine when smoke bombs are readily available around Fourth of July! Use care when handling or lighting any smoke bomb, and only do this activity outdoors. Start by placing the smoke bomb on a flat surface and light it. Use the hole in the bottom of the trash can to cover

the smoke bomb and fill the can with smoke. Use caution as hot debris can shoot up from the smoke bomb and burn tiny holes in the plastic membrane. The trash can is ready for you to tap the membrane and produce dozens of colorful smoke rings.

Whether you choose the more traditional smoke machine or the much more exciting smoke bomb, the rule is the same—never shoot smoke in anyone's face (people or animals)! Aim the flying rings of smoke in the air and fire away. I also recommend that you do this activity outside, or be prepared for the smoke alarms to go off. It's always interesting to have to explain the smoke rings to the fire department, especially when the rings are created by smoke bombs . . . been there, done that.

WHAT'S GOING ON HERE?

The proper name for the air cannon device is **vortex generator**. The blast of air that shoots out of the cannon is actually a flat vortex of air, similar to rings of smoke blown by a talented cigar smoker. Please note that this is not an endorsement of talented cigar smokers, nor should you ever take up smoking, but I think you know what I mean. Wow . . . I'll think twice before ever mentioning talented cigar smokers again.

A vortex is generated because the air exiting the container at the center of the hole is traveling faster than the air exiting around the edge of the hole. Bernoulli's Principle states that the faster a flow of air is moving the lower its pressure. Since the air inside the vortex is moving faster than the outside air, the resulting inward pressure is the force that holds the smoke ring together. Eventually, air friction steals away all the energy stored in the vortex and the smoke ring drifts to a stop. Very cool!

This activity demonstrates the simple concept that air occupies space. Bernoulli really understood the idea that fast-moving air creates an area of low pressure . . . and the flying smoke rings are an added bonus.

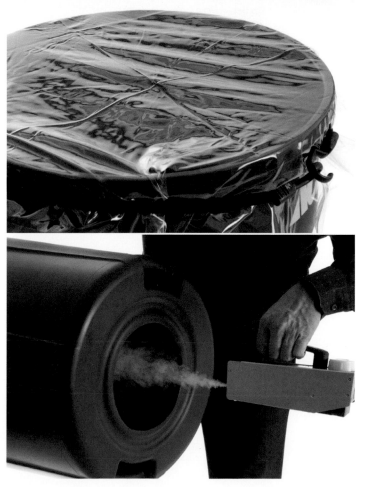

KITCHEN
CHEMISTRY

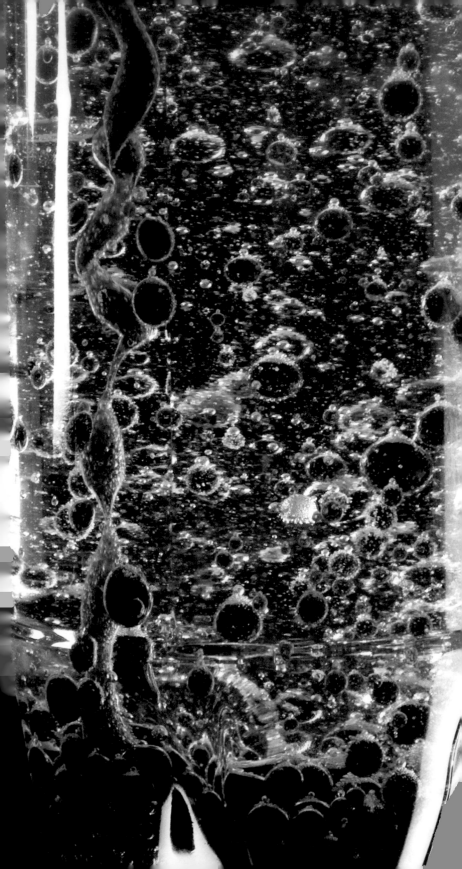

BUBBLING **LAVA BOTTLE**

Learn how to make a wave bottle using oil, water, and a secret ingredient
that makes the whole thing fizz, bubble, and erupt.

WHAT YOU NEED

Clean, plastic soda bottle
with a cap (16-ounce
size works well)

Vegetable oil (the
cheaper the better)

Food coloring

Alka-Seltzer tablet

Large flashlight

Water

LET'S TRY IT!

1. Fill the bottle three-quarters full with vegetable oil.

2. Fill the rest of the bottle with water (almost to the top but not overflowing).

3. Add about ten drops of food coloring. Be sure to make the water fairly dark in color. Notice that the food coloring only colors the water and not the oil.

4. Divide the Alka-Seltzer tablet into four pieces.

5. Drop one of the tiny pieces of Alka-Seltzer into the oil and water mixture. Watch what happens. When the bubbling stops, add another chunk of Alka-Seltzer.

6. When you have used up all of the Alka-Seltzer and the bubbling has completely stopped, screw on the soda bottle cap. Tip the bottle back and forth and watch the wave appear. The tiny droplets of liquid join together to make one big wavelike blob.

TAKE IT FURTHER

As you watched the bubbling color blobs rise and fall in the water, you probably thought to yourself, "This is just like a lava lamp . . . without the lamp!" On a side note, if you have no concept of what a lava lamp is, pull out your smartphone and Google it.

How to Make a Lava Lamp

To make a cool looking lava lamp, you'll need a large flashlight like the one in the photograph. Carefully rest the bottle of oil and water directly on the lens of the flashlight and repeat the experiment above with the bright light shining up and through the liquid. Groovy, baby!

WHAT'S GOING ON HERE?

First of all, you confirmed what you already know—oil and water do not mix. Even if you try to shake up the bottle, the oil breaks up into small little drops, but it doesn't mix with the water. Why is it that oil and water are such opposites?

Oil and water don't mix because of how their molecules are constructed. Water is what is known as a **polar molecule**. A water molecule is shaped like a V, with an oxygen atom at the bottom point of the V and a hydrogen atom on each of the two top ends. However, there is unequal sharing of electrons between the hydrogen and oxygen atoms. This means that the bottom of the molecule has a negative electrical charge, while the top carries a positive charge.

Vegetable oil, on the other hand, is a **nonpolar molecule** made of long chains of hydrocarbons—strings of carbon atoms bonded to hydrogen atoms. Unlike the water molecule, there is equal sharing of electrons between the carbon and hydrogen atoms. This means that the electrical charges of the atoms are not separated, so the molecules don't have opposite positive and negative ends.

If you were to think of molecules like groups of people, the polar molecules hang out with other polar molecules, and the nonpolar molecules with other nonpolar molecules. This brings us back to the reason why oil and water don't mix. Water is a polar molecule, and it just doesn't hang out with nonpolar molecules like oil. Scientists say that oil and water are **immiscible**.

The adage "like dissolves like" will help you remember what will mix with what. Salt and water mix because both molecules are polar—like dissolves like. It's also easy to mix vegetable oil and olive oil, or motor oil and peanut oil . . . but that's gross.

You also noticed that food coloring only mixes with water . . . and now you know why. Food coloring is a polar molecule because it dissolves in water. In other words, food coloring and water are **miscible**. Vegetable oil is not affected by the food coloring because they are polar opposites.

Here's the surprising part . . . the Alka-Seltzer tablet reacts with the water to make tiny bubbles of carbon dioxide. These bubbles attach themselves to the blobs of colored water and cause them to float to the surface. When the bubbles pop, the color blobs sink back to the bottom of the bottle, and the whole thing starts over until the Alka-Seltzer is used up. When the chemical reaction between the Alka-Seltzer and water is over and the bubbling stops, you're left with a cool looking wave bottle that will sit proudly on your desk.

COLOR CHANGING **MILK**

This is guaranteed to become one of your favorite kitchen chemistry experiments. Some very unusual interactions take place when you mix a little milk, food coloring, and a drop of liquid soap. Use this experiment to amaze your friends and uncover the scientific secrets of soap.

WHAT YOU NEED

Milk (whole or 2%)

Dinner plate

Food coloring (red, yellow, green, blue)

Dish-washing soap (Dawn brand works well)

Cotton swabs

LET'S TRY IT!

1. Pour enough milk in the dinner plate to completely cover the bottom to the depth of about ¼ inch. Allow the milk to settle before moving on to the next step.

2. Add one drop of each of the four colors of food coloring—red, yellow, green, and blue—to the milk. Keep the drops close together in the center of the plate of milk.

3. Find a clean cotton swab for the next part of the experiment. Predict what will happen when you touch the tip of the cotton swab to the center of the milk. It's important not to stir the mix—just touch it with the tip of the cotton swab. Go ahead and try it.

4. Now place a drop of liquid dish soap on the other end of the cotton swab. Place the soapy end of the cotton swab back in the middle of the milk and hold it there for 10 to 15 seconds. Look at that burst of color! It's like the Fourth of July in a plate of milk.

5. Add another drop of soap to the tip of the cotton swab and try it again. Experiment with placing the cotton swab at different places in the milk. Notice that the colors in the milk continue to move even when the cotton swab is removed. What makes the food coloring in the milk move?

TAKE IT FURTHER

Repeat the experiment using water in place of milk. Will you get the same eruption of color? What kind of milk produces the best swirling of color, skim, 1%, 2%, or whole milk? Why? This is the basis of a great science fair project as you compare the effect that the dishwashing soap has on a number of different liquids. Do you see any pattern in your observations?

WHAT'S GOING ON HERE?

Milk is mostly water, but it also contains vitamins, minerals, proteins, and tiny droplets of fat suspended in solution. Fats and proteins are sensitive to changes in the surrounding solution (the milk).

The secret of the bursting colors is in the chemistry of that tiny drop of soap. Dish soap, because of its bipolar characteristics (nonpolar on one end and polar on the other), weakens the chemical bonds that hold the proteins and fats in solution. The soap's nonpolar, or **hydrophilic** (water-loving), end dissolves in water, and its **hydrophobic** (water-fearing) end attaches to a fat globule in the milk. This is when the fun begins.

The molecules of fat bend, roll, twist, and contort in all directions as the soap molecules race around to join up with the fat molecules. During all of this fat molecule gymnastics, the food coloring molecules are bumped and shoved everywhere, providing an easy way to observe all the invisible activity. As the soap becomes evenly mixed with the milk, the action slows down and eventually stops. This is why milk with a higher fat content produces a better explosion of color—there's just more fat to combine with all of those soap molecules.

Try adding another drop of soap to see if there's any more movement. If so, you discovered there are still more fat molecules that haven't found a partner at the big color dance. Add another drop of soap to start the process again.

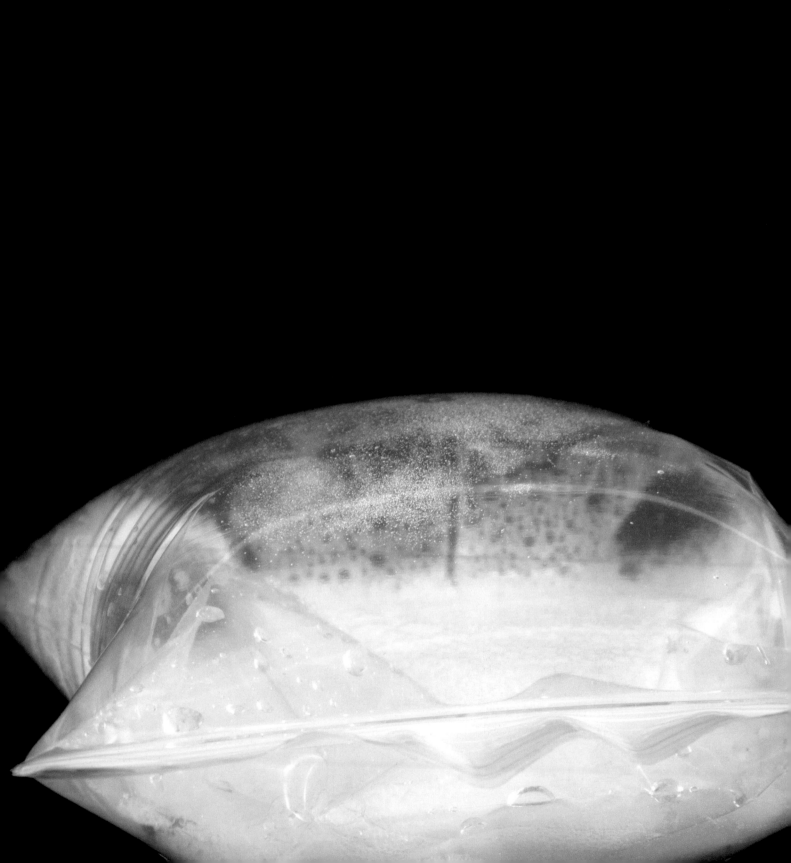

HOW TO MAKE A CO₂ SANDWICH

Mom always warned us never to play with our food, but no one said that the wrappers were off-limits. Here's a fun activity that uses some common items you'll find around the house and a little creativity to explore the "pop" factor of vinegar and baking soda.

WHAT YOU NEED

Safety glasses

Measuring cup and spoons

Vinegar

Baking soda

Quart-size zipper-lock bags and snack-size zipper-lock bags

WARNING! IMPORTANT SAFETY RULES

This activity requires safety glasses and adult supervision. Why? Because I said so. Isn't that something else that Mom always said?

LET'S TRY IT!

1. Start by putting on your safety glasses.
2. Fill three quart-size zipper-lock bags with approximately 1 tablespoon of baking soda.
3. Fill three smaller snack-size zipper-lock bags with varying amounts of vinegar. For example, fill one bag with 60 mL (¼ cup) of vinegar, the next bag with 80 mL (⅓ cup) of vinegar, and the last bag with 120 mL (½ cup) of vinegar.
4. Seal the vinegar bags and place them inside the bags with the baking soda. When you seal the outside bags, make sure to remove as much of the air as possible.
5. Put the bags on a table where it's okay for things to get a little wet and messy (outside tables would be good).
6. Now get ready for the fun . . . The goal is to break open the smaller bag filled with vinegar in order for it to mix with the baking soda. One way to bust open the bag is to smack your fist down on the vinegar bags inside the baking soda bags to break them open. Immediately shake the bags to make sure the substances mix.
7. Make observations about how large each bag gets and how long it takes before you hear the giant POP!

TAKE IT FURTHER

What would happen if you diluted the vinegar with a little water? How would this affect the expanding gas? Predict what would happen if you used warm water instead of cold water to dilute the vinegar.

Try changing the amount of vinegar or baking soda you use to see how the reaction changes. Remember to change only one variable at a time. For example, you can increase or decrease the amount of vinegar, but be sure to keep the amount of baking soda (1 tablespoon) the same. Likewise, you can keep the amount of vinegar the same but change the amount of baking soda in each bag. By changing only one variable at a time, you'll be able to determine which ingredient has the most impact on the POP! What are three more variables you can change?

Here's another cool twist on the experiment . . . Pour 4 tablespoons of vinegar into a clean empty bottle. Carefully drop 1 tablespoon of baking soda into the neck of a latex balloon. Shake the balloon to make sure the baking soda falls all the way into the tip of the balloon. Stretch the neck of the balloon over the top of the bottle and gently lift the balloon, making sure that the baking soda drops down into the bottle. As the baking soda reacts with the vinegar, the balloon inflates on its own.

WHAT'S GOING ON HERE?

Sure, bubbling liquids and popping bags are fun, but what's the science behind the exploding sandwich bag? When you mix vinegar and baking soda, a chemical reaction takes place producing a gas called carbon dioxide.

The reaction that happens from mixing vinegar and baking soda is caused by the chemical reaction between the acetic acid (CH_3COOH) in vinegar and the sodium bicarbonate ($NaHCO_3$) in baking soda. This reaction forms sodium acetate ($NaCH_3COO$), water (H_2O), and carbon dioxide (CO_2).

The chemical equation is as follows:

$$CH_3COOH + NaHCO_3 \rightarrow NaCH_3COO + H_2O + CO_2$$

The bag puffs up because the CO_2 takes up lots of space, eventually filling the bag. If there's more gas than the bag can hold . . . KABOOM! If you're lucky, the zipper-lock seal will bust open, but the bag will not break. Now you can reuse the bag to make another CO_2 sandwich. Separating the substances in bags is a clever way of slowing down the reaction.

SOAP **SOUFFLÉ**

Ivory soap . . . it's the soap that floats. But why? Discover the secret behind this floating sensation by cooking the whole bar of soap in the microwave. That's right, a bar of Ivory soap + the microwave oven = a very cool trick! And your kitchen will smell so fresh and clean when you're finished.

WARNING! IMPORTANT SAFETY RULES

This experiment requires the use of a microwave oven. Even though you've used a microwave oven to heat a leftover burrito a thousand times, the lawyers make us include this warning—adult supervision is required.

LET'S TRY IT!

1. The first part of this experiment is designed to prove whether or not the claim is true. Does Ivory soap really float? Fill the bowl with water and drop in a brand-new bar of Ivory soap. It's a pretty simple test . . . does it float?

2. Maybe all bars of soap float. If you have other brands of soap, try the float or sink test. You'll probably discover that all of the bars of soap sink except for the Ivory brand soap. Why?

3. Remove the Ivory soap from the water and break it in half to see if the bar of soap is actually hollow or if there are huge pockets of air. If either is true, that would make the soap float, right?

4. Use the knife to cut the bar of Ivory soap into four equal pieces. Place the pieces of soap on a dinner plate, and then place the whole thing in the center of the microwave oven, after asking permission from an adult.

5. Cook the bar of soap on HIGH for 1 minute. Don't take your eyes off the bar of soap as it begins to expand and erupt into beautiful puffy clouds. Be careful not to overcook your soap soufflé.

6. Allow the soap to cool for a minute or so before touching it. Amazing. . . it's puffy but rigid. Don't waste the soap. Take it into the shower or bath. It's still great soap with a slightly different shape and size.

TAKE IT FURTHER

Try the same experiment with any bar of soap other than Ivory. Do you see the same results? If you have an older bar of Ivory soap around the house, do a side-by-side comparison test between the older soap and a brand-new bar from the store. Does the age of the soap have any effect on the size of the soap soufflé?

What would happen if you microwaved a marshmallow? What are you waiting for? Try it! Put the marshmallow on a plate and microwave it for about 30 seconds. What happens? Run the test several other times with new marshmallows and different "cooking" times.

WHAT'S GOING ON HERE?

Ivory soap is one of the few brands of bar soap that floats in water. But when you break the bar of soap into several pieces, there are no large pockets of air inside. If it floats in water and has no large pockets of air, it must mean that the soap itself is less dense than water. Ivory soap floats because air is whipped into the soap during the manufacturing process. If you break the bar of soap in half with your hands and look closely at the edge of the bar, you'll see tiny pockets of air. Cutting the soap with a knife leaves a smooth edge making it impossible to see the exposed air bubbles.

The air-filled soap was actually discovered by accident in 1890 by an employee at Procter & Gamble. While mixing up a batch of soap, the employee forgot to turn off his mixing machine before taking his lunch break. This caused so much air to be whipped into the soap that the batch nearly doubled in size. When the soap was formed into bars, the bars floated in water. The response by the public was so favorable that Procter & Gamble continued to whip air into the soap, capitalizing on the mistake by marketing their new creation as "The Soap that Floats!"

Why does the soap expand in the microwave? This is actually very similar to what happens when popcorn pops or when you try to microwave a marshmallow. Those air bubbles in the soap (or in the popcorn kernels or marshmallow) contain water molecules. Water is also caught up in the matrix of the soap itself. The expanding effect is caused when the water is heated by the microwave. The water vaporizes and the heat causes the trapped air to expand. Likewise, the heat causes the soap itself to soften and become pliable.

This effect is actually a demonstration of **Charles's Law**. Charles's Law states that as the temperature of a gas increases, so does its volume. When the soap is heated, the molecules of air in the soap move quickly, causing them to move far away from each other. This causes the soap to puff up and expand to an enormous size. Other brands of soap without whipped air tend to melt when heated up in the microwave.

And now the entire kitchen smells like . . . cooked soap.

RED CABBAGE **CHEMISTRY**

Ahhh, the sweet smell of science! Invite your friends over to share in this super smelly but really cool activity. Plug your nose and get ready to make your own red cabbage indicator that will test the acidity or alkalinity of certain liquids.

WHAT YOU NEED

Red cabbage

Blender

Strainer

Clear drinking glasses

White paper

Apron or lab coat
(cabbage juice can
leave nasty stains!)

Test chemicals:
Vinegar, baking soda,
lemon juice, washing soda,
laundry detergent, soda
pop, and Alka-Seltzer

LET'S TRY IT!

1. Peel off three or four big cabbage leaves and put them in a blender filled one-half full with water. Blend the mixture on high until you have purple cabbage juice.

2. Pour the purplish cabbage liquid through a strainer to filter out all of the big chunks of cabbage. Doesn't cabbage juice smell great? Save the liquid for the experiments to follow.

3. Set out three glasses, side by side, against a white piece of paper as the background. Fill each glass one-half full with cabbage juice.

4. Add a little vinegar to the first glass of cabbage juice. Stir with a spoon and notice the color change to red, which indicates that vinegar is classified as an acid.

5. In the second glass, add a teaspoon of washing soda or laundry detergent. Notice how the liquid turns green, indicating that this chemical is a base. Keep these two glasses of red and green liquid for future reference, along with the third glass of purple cabbage juice to show the color of a neutral solution.

6. Fill additional glasses with purple cabbage juice. Try adding each of the other "test chemicals" to a small amount of cabbage juice and note the color change to determine if the chemical is an acid or a base.

TAKE IT FURTHER

Use your cabbage juice indicator to test the acid or base properties of other common substances. You might want to try orange juice, lemonade, milk, salt, ammonia, or soap.

You can also make your own pH indicator strips, like you see lifeguards using to test the pH of pool water (their indicator strips are not made from red cabbage so you probably shouldn't dip your cabbage-soaked strips into the pool). Soak some coffee filter paper in concentrated cabbage juice. Remove the paper from the cabbage juice and hang it up by a clothespin to dry. Cut the dried paper into thin strips. Dip the strips into various liquids to test their pH. The redder the strip turns, the more acidic the liquid is. The greener the strip turns, the more basic the liquid is.

WHAT'S GOING ON HERE?

Some substances are classified as either an acid or a base. Think of acids and bases as opposites—acids have a low pH and bases have a high pH. For reference, water (a neutral) has a pH of 7 on a scale of 0–14. Scientists can tell if a substance is an acid or a base by means of an **indicator**. An indicator is typically a chemical that changes color if it comes in contact with an acid or a base.

As you can see, the purple cabbage juice turns red when it mixes with something acidic and turns green when it mixes with something basic. Red cabbage juice is considered to be an indicator because it shows us something about the chemical composition of other substances.

What is it about cabbage that causes this to happen? Red cabbage contains a water-soluble pigment called **anthocyanin** that changes color when it is mixed with an acid or a base. The pigment turns red in acidic environments with a pH less than 7 and the pigment turns bluish-green in alkaline (basic) environments with a pH greater than 7.

Red cabbage is just one of many indicators that are available to scientists. Some indicators start out colorless and turn blue or pink, for example, when they mix with a base. If there is no color change at all, the substance that you are testing is probably neutral, just like water.

EATING NAILS FOR BREAKFAST

Have you ever taken the time to read the nutritional information on your box of breakfast cereal? You'll find that your cereal contains more than just wheat and corn. In fact, you'll notice that your cereal contains sodium, potassium, calcium, and iron . . . iron? Some nails are made from iron. Could you be eating nails for breakfast? Well, not really, but certain cereals do have a very high iron content. To better explain this, try the following experiment.

LET'S TRY IT!

1. Open the box of cereal and pour a small pile of flakes on the plate. Crush them into tiny pieces with your fingers. Spread out the pile so it forms a single layer of crumbs on the plate. Bring the magnet close to the layer of crumbs (but don't touch any) and see if you can get any of the pieces to move. Take your time. If you get a piece to move without touching it, that piece may contain some metallic iron.

2. Firmly press the magnet directly onto the crumbs but don't move it. Lift it up and look underneath to see if anything is clinging to the magnet. Several little pieces may be stuck there. Is it the magnet being attracted to static electricity or just sticky cereal? It could be the iron. Throw away the small pile of cereal and clean off your magnet in order to move on to the next step.

3. Pour a little water onto the plate and float a few large flakes on the surface. Hold the magnet close to (but not touching) a flake and see if the flake moves toward the magnet. (The movement may be very slight, so be patient.) With

Magnets come in all shapes, sizes, and strengths. Ask at your local hardware store for a strong magnet for a science experiment. The strongest magnets in the world are called neodymium, or "rare-earth," magnets. They are ten times stronger than standard ceramic magnets and are commonly used in speakers and computer disc drives. It is possible to pull the iron out of cereal using a standard magnet, but you'll get much better results using a neodymium magnet.

practice, you can pull the flakes across the water, spin them, and even link them together in a chain. Hmm . . . there must be something that's responding to the magnet. Could it be metallic iron? In your cereal?

4. It's time to mix up a batch of cereal soup to further investigate the claim of iron in your breakfast cereal. Open a quart-size zipper-lock bag and measure 1 cup of cereal (that's equal to one serving according to the nutritional information on the side of the cereal box) into the empty bag. Fill the bag one-half full with warm water and carefully seal it, leaving an air pocket inside.

5. Give the cereal and water a good mixing by shaking the bag around for a minute or so. After a few minutes, the warm water will start to dissolve the flakes of cereal and the liquid will turn into a brown, soupy mixture. Allow the mixture to sit for at least 20 minutes before moving on to the next step.

6. Make sure the bag is tightly sealed and position it on a flat surface in the palm of your hand. Place the strong magnet on top of the bag. Put your other hand on top of the magnet and flip the whole thing over so the magnet is underneath the bag. Slowly slosh the contents of the bag in a circular motion for 15 or 20 seconds. The idea is to attract any free-moving bits of metallic iron in the cereal to the magnet.

7. Use both hands again and flip the bag and magnet over so the magnet is on top. Gently squeeze the bag to raise the magnet a little above the cereal soup. Don't move the magnet just yet. Look closely at the edges of the magnet where it's touching the bag. You should be able to see tiny black specks on the inside of the bag around the edges of the magnet. That's the iron!

8. Keep one end of the magnet touching the bag and move it in little circles. As you do this, the iron will gather into a bigger clump and become much easier to see. Few people have ever noticed iron in their food, so you can really impress your friends with this one. When you're finished, simply pour the soup down the drain and rinse the bag.

TAKE IT FURTHER

By this time, your brain should be overflowing with questions. Is that black stuff really iron? Take a trip to the grocery store to investigate the contents of other cereals. What other brands claim to have iron? Conduct the same experiment using other brands of cereal to see if you can find more magnetic black stuff.

WHAT'S GOING ON HERE?

Many breakfast cereals are fortified with food-grade iron (chemical symbol: Fe) as a mineral supplement. Metallic iron is digested in the stomach and eventually absorbed in the small intestine. If all of the iron from your body were extracted, you'd have enough iron to make two small nails.

Iron is found in a very important component of blood called hemoglobin. Hemoglobin is the compound in red blood cells that carries oxygen from the lungs so it can be utilized by the body. It's the iron in the hemoglobin that gives blood its red appearance.

A diet without enough iron can cause you to be tired, catch diseases more easily, and make your heart and breathing rates too fast. Food scientists say that a healthy adult requires about 18 mg of iron each day. As you can see, iron plays a very important part in maintaining a healthy body. Score! Cereal for dinner!

TACO SAUCE PENNY CLEANER

It's one of those things you hear about but wonder if it's true. Can you really use taco sauce to clean the tarnish off of a penny? Surprisingly, taco sauce does a great job of cleaning pennies, but how does it work?

WHAT YOU NEED

Dirty pennies (try to collect tarnished pennies that all look the same)

Taco sauce

Vinegar

Tomato paste

Salt

Water

Small plates

Masking tape or sticky note

LET'S TRY IT!

1. Let's start with the hypothesis of this experiment: Taco sauce is a great penny cleaner. If this is true, then we can use the scientific method to determine the science behind this saucy secret.

2. Place several tarnished pennies on a plate and cover them with taco sauce. Use your fingers to smear the taco sauce all over the top surfaces of the pennies. Remember to wash your hands . . . and don't lick your fingers because dirty pennies are gross.

3. Allow the taco sauce to sit on the pennies for at least 2 minutes.

4. Rinse the pennies in the sink and look at the difference between the top sides of the pennies that touched the taco sauce and the bottom sides. It's no myth—taco sauce does the trick.

5. For an easy-to-see comparison, use another tarnished penny and cover only half of the surface of the penny with taco sauce. Don't smear the sauce around with your finger this time—you want a nice dividing line between the two sides. Let the penny and sauce combo sit for a few minutes and rinse. It's a cool half-and-half penny.

Since the hypothesis is true, which of the ingredients in taco sauce is responsible for its cleaning power? Let's find out . . .

1. The list of ingredients on the packet of taco sauce reveals four substances to test: vinegar, tomato paste, salt, and water.

2. Place two equally tarnished pennies on each of four different plates. Use masking tape or a sticky note to mark each plate with the taco sauce ingredient you are testing (vinegar, tomato paste, salt, or water).

tomato
paste

tomato
paste

Vinegar

3. Cover the pennies with the various ingredients, smear them around with your fingers, and allow the pennies to sit for at least 2 minutes. Be sure to wash your hands.

4. Rinse the pennies from each test plate with water. Which ingredient cleaned the pennies the best?

None of the individual ingredients do a good job of cleaning the dirty pennies. In fact, the results are less than impressive. Maybe two or more of the ingredients work together to react against the copper oxide (the tarnish) on the penny. Let's find out . . .

1. Place two equally tarnished pennies on each of three different plates. Make three signs that say "Tomato Paste + Vinegar," "Salt + Vinegar," and "Tomato Paste + Salt."

2. Cover the pennies with each of the mixtures, smear them around with your fingers, and give the ingredients at least 2 minutes to react. Wash your hands.

3. Rinse the pennies under water. Now what do you notice?

4. And the winner is . . . VINEGAR + SALT! But why?

WHAT'S GOING ON HERE?

The clear winner is the mixture of vinegar and salt. Neither vinegar nor salt by itself cleaned the pennies, but when they were mixed together something happened. The chemistry behind the reaction is somewhat complicated but very interesting. When the salt and the vinegar are mixed together, the salt dissolves in the vinegar solution and breaks down into sodium and chloride ions. The chloride ions then combine with the copper in the penny to remove the tarnish or copper oxide from the surface of the penny. It is also well known that a mixture of lemon juice and salt does a good job of removing tarnish from metals and works very well on pennies. By themselves, the salt and vinegar do very little in the way of removing the coating of copper oxide on the penny, but together these ingredients make a great cleaning agent.

So, the secret in taco sauce is the combination of the ingredients. Someone might argue that tomato paste is slightly acidic and may contribute in a small way to removing the copper oxide coating, but the real "power ingredients" are salt and vinegar.

SCIENCE FAIR CONNECTION

The "Taco Sauce Penny Cleaner" is a great example of a science fair project. First, you ask a question—does taco sauce really clean pennies? You find that it does

and then you ask another question—what is it in the taco sauce that causes it to clean pennies? You run multiple tests and isolate one variable at a time to see if the vinegar, the tomato paste, the salt, or the water is the real cleaning agent for the pennies. Guess what? Nothing cleans the penny. Now what do you do? You ask another question—could a combination of ingredients cause the cleaning action? Again, you isolate the variables and eventually reach the conclusion that a combination of vinegar and salt cleans the pennies.

The "Taco Sauce Penny Cleaner" experiment clearly shows **scientific inquiry** in motion—ask a question, run some tests, ask another question, run some tests, ask another question, run some tests, and eventually come to some conclusions. Good science fair projects should leave you with more questions than they answer. What do you still wonder about? How could you extend the experiment to try to find some more answers? Did this activity cause to you wonder about something else entirely? Could you create a new experiment based on your new questions?

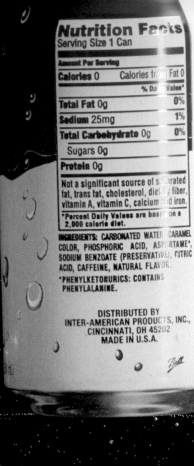

SINKING SODA SURPRISE

I remember as a kid diving to the bottom of the swimming pool to retrieve unopened cans of soda pop. No, someone didn't accidentally drop the cans of soda into the pool. Instead, our parents tossed the unopened cans of soda into the pool as a game on the Fourth of July or any time they wanted to keep us busy. Come to think of it, I'm not sure why we kept retrieving the cans, but we did. So, based on my personal experience, I knew that unopened cans of soda sink in water . . . until I saw my first can of soda float. Why do some cans of soda float and others sink?

WHAT YOU NEED

An assortment of unopened soda cans (diet, regular, brand name, generic)

Note: Use standard 12-ounce cans. Mini-cans will not work.

A large, deep container of water like a 5-gallon bucket or an aquarium

Note: If the can of regular soda floats, there might be an air bubble trapped under the bottom of the can. Tip the can to the side to release the pocket of air trapped under the can.

LET'S TRY IT!

1. Ask your audience the question: "Will this can of regular soda float or sink in the bucket of water?" After gathering everyone's answer, place the can of regular soda in the water and notice that it sinks to the bottom.

2. Pick up a can of diet soda and pose the same question. Be sure to point out the fact that the cans are exactly the same size and shape and contain the same amount of liquid (compare the number of milliliters—probably 355 mL). Place the can of diet soda in the water. It floats! Wobble the can from side to side to show your audience that there are no bubbles trapped under the bottom. It still floats. Why?

3. Let your group experiment with different kinds of soda. Why do the diet sodas float and the regular sodas sink, no matter the brand?

TAKE IT FURTHER

Try the experiment again using salt water. Are your results any different? What if you continue adding salt? How much salt do you have to add before your results change?

Consider changing the temperature of the water or the temperature of the cans. Do either of those changes affect the results?

WHAT'S GOING ON HERE?

This demonstration is an excellent way to learn about density. You can determine how dense an object is by using the density equation: Density = Mass ÷ Volume. The density of water is 1.0 grams per milliliter. Any object that has a density greater than 1.0 g/mL will sink and any object with a density of less than 1.0 g/ mL will float.

Let's look at the variables. **Volume** refers to how much space an object occupies. For fluids, volume is usually measured in liters (L) or milliliters (mL). When

you compared the different cans of soda, you probably noticed that the cans contained the same amount of liquid (355 mL). Since the cans of soda used for this experiment have the same volume, the only variable left to consider in the density equation is the mass. In order for the density to change, the mass must change . . . and it does!

If a diet and a regular soda can are placed on a double-pan balance scale, it quickly becomes clear that the regular soda is heavier (or has more mass) than the diet soda. Mass refers to how much stuff exists within an object, and for the purposes of our experiment, mass is measured in grams.

What's inside the can of regular soda that makes it so much heavier than diet soda? Comparing the list of ingredients on both cans you see that a can of regular soda has about 150 calories while diet soda has 0 calories. All of those calories come from one place . . . sugar! Diet sodas usually contain aspartame, an artificial sweetener, while regular sodas use sugar. Take a look at the nutritional information on the side of the cans.

Notice how much sugar is in a regular soda (look under carbohydrates). Most regular sodas have between 39 and 43 grams of sugar. This added mass is why the cans of regular soda sink in water.

So, how much is 40 grams of sugar? If you go to a restaurant and order tea, some people add two or three packets of sugar (okay, some people stir in six packets of sugar, but we're talking about the average person). How many packets of sugar make up 40 grams? Of course, not every packet of sugar is the same, but a little research of the published nutritional information from fast food restaurants in the United States revealed that a packet contains about 4 grams of sugar, and those 4 grams of sugar have 15 calories. This means that there are about ten packets of sugar in a can of regular soda, accounting for the 150 calories. That's a lot of sugar!

So the density of a soda actually depends on how much sugar or sweetener is used. The 40 grams of sugar added to a can of regular soda make it sink, and the relatively tiny amount of artificial sweetener used in diet soda has a negligible effect on the mass, enabling the can to float.

SEVEN-LAYER **DENSITY COLUMN**

Anyone can stack blocks, boxes, or books, but only those with a steady hand and a little understanding of chemistry can stack liquids. What if you could stack seven different liquids in seven different layers? Think of it as a science burrito.

LET'S TRY IT!

1. Measure 8 ounces of each type of liquid into the seven plastic cups. Depending on the size of the glass cylinder, you might need more or less of each liquid—8 ounces is just a good starting point. You may want to color the corn syrup and the rubbing alcohol with a few drops of food coloring to create a more dramatic effect in your column. Here is the order of layers starting from the bottom and working your way to the top:

> **Honey**
> **Corn syrup**
> **Dish soap**
> **Water**
> **Vegetable oil**
> **Rubbing alcohol**
> **Lamp oil**

2. Start your column by pouring the honey into the cylinder. It is very important to pour the liquids carefully into the center of the cylinder. Make sure the honey does not touch the sides of the cylinder while you are pouring. It's important to let each layer settle before adding the next one. Take your time and pour slowly and carefully.

3. The next layer is corn syrup. Again, try not to let the corn syrup touch the sides of the container as you're pouring. The key is to pour slowly and evenly.

4. Repeat the same procedure with the dish soap. Pour the soap directly into the middle of the layer of corn syrup . . . and take your time pouring!

5. Stop for just a second to enjoy your success. You're almost halfway to your goal of stacking seven layers of liquid. The next liquid is water, and you'll need to use the food baster—it's like a giant medicine dropper for food. From this point forward, it's okay to let the liquids touch the sides of the cylinder. In fact, it's a must! Dip the tip of the food baster in the cup of water, squeeze the bulb, and draw up some water. Rest the tip of the food baster on the inside wall of the cylinder and slowly squeeze the bulb. Let the water slowly trickle down the glass to create the next layer. Take your time!

6. You'll use the food baster once again for the next layer—vegetable oil. Use the inside wall of the cylinder to let the vegetable oil slowly trickle down and form the next layer.

7. Wash the food baster with some soap and water in the sink before moving on to the rubbing alcohol. If you have not already colored the rubbing alcohol, use a couple drops of food coloring to make sure this layer isn't confused with water. Use the food baster and the inside wall of the cylinder to add this next layer.

8. You're one layer away from success. Again, rinse the food baster in the sink before moving on to the lamp oil. Since lamp oil is flammable, you must do this last step away from any open flames. Use the food baster to draw up some lamp oil, which has a low surface tension and easily leaks out of the food baster. Keep your finger over the tip as you transport it over to the cylinder. By now you're a pro at this. Use the baster and the inside wall of the cylinder to slowly add the final liquid layer.

9. Take your much-deserved bow and accolades from the guests in the viewing stands (or your friends hanging out in the kitchen). You've made a seven-layer science burrito, so to speak.

TAKE IT FURTHER

If you want to create an even cooler science burrito, add the "meat and black olives." In other words, select a few items from around the house (safety pin, key, staple,

peanut, raisin, chocolate chip, small rubber bouncy ball, ping pong ball, etc.—be creative!) and carefully drop each item individually into the center of the cylinder. Some items will stay on or near the top of the stack of liquids and other items will sink part or all of the way down to the bottom of the cylinder. Why the difference? The densities and masses of the objects you drop into the liquids vary. If the layer of liquid is more dense than the object itself, the object stays on top of that liquid. If the layer of liquid is less dense than the object, the object sinks through that layer until it meets a liquid layer that is dense enough to hold it up.

Here's something else you can do to illustrate the connection between weight (or mass) and density. Set up a scale and weigh each of the liquids from your column. Make sure that you weigh equal portions of each liquid. You should find that the weights of the liquids correspond to their level in the column. For example, the honey will weigh more than the corn syrup. By weighing these liquids, you will find that density and weight are closely related.

WHAT'S GOING ON HERE?

The science secret here is **density**. Density is a measure of how much mass is contained in a given unit volume (Density = Mass ÷ Volume). If mass is a measure of how much "stuff" there is in an object or liquid, density is a measure of how tightly that "stuff" is packed together.

Based on this density equation (Density = Mass ÷ Volume), if the weight (or mass) of something increases but the volume stays the same, the density has to go up. Likewise, if the mass decreases but the volume stays the same, the density has to go down. Lighter liquids (like water or rubbing alcohol) are less dense or have less "stuff" packed into them than heavier liquids (like honey or corn syrup).

Every liquid has a density number associated with it. Water, for example, has a density of 1.0 g/cm^3 (grams per cubic centimeter—another way to say this is g/mL, which is grams per milliliter). Here are the densities of the liquids used in the column, as well as other common liquids:

MATERIAL	DENSITY (g/cm^3 or g/mL)
Rubbing alcohol	0.79
Lamp oil (refined kerosene)	0.81
Baby oil	0.83
Vegetable oil	0.92
Ice cube	0.92
Water	1.00
Milk	1.03
Dawn dish soap	1.06
Light corn syrup	1.33
Maple syrup	1.37
Honey	1.42

The numbers in the table are based on data from manufacturers of each item. Densities may vary from brand to brand. You'll notice that according to the number, rubbing alcohol should float on top of the lamp oil, but we know from our experiment that the lamp oil is the top layer. Chemically speaking, lamp oil is nothing more than refined kerosene with coloring and fragrance added. Does every brand of lamp oil exhibit the same characteristics? Sounds like the foundation of a great science fair project.

So, the next time you're enjoying a glass of iced tea, you'll know why those ice cubes float. That's right . . . it's all about density.

Lamp Oil

Rubbing Alcohol

Vegi Oil

Water

Dawn Dish Soap

Karo Syrup

Honey

EGG IN THE BOTTLE TRICK . . .
WITH A MODERN-DAY TWIST

Here's a classic science experiment that is more than a hundred years old and is guaranteed to amaze your friends. The original demonstration used a hard-boiled egg and a glass milk bottle. Since old milk bottles are hard to come by, here's a modern-day version of the same experiment using a juice bottle and a water balloon. You'll also get a chance to try your hand at getting a real egg into a juice bottle using a cool, upside down twist.

WHAT YOU NEED

A wide-mouth, glass juice bottle

Peeled hard-boiled eggs (just a little wider in size than the opening of the glass bottle)

Several strips of paper (2 x 6 inches)

Matches

Balloons (9-inch balloons work great)

Water

WARNING! IMPORTANT SAFETY RULES
This activity requires the use of matches and fire. Adult supervision is required.

LET'S TRY IT!

1. Carefully fill the balloon with water so the balloon is slightly larger than the mouth of the bottle. Tie it off. Make several water balloons just in case the first one breaks.

2. The glass juice bottle should have a wide mouth between 1½ and 2 inches in diameter. If you can find an old-fashioned glass milk bottle, use it! Rinse out the bottle to remove any leftover, sticky, slimy stuff that might be at the bottom.

3. Here's the challenge . . . Your job is to find a way to get the balloon into the bottle without breaking it. How are you going to do it? It's important that you take a minute to test out some of your ideas before jumping ahead to read our solution. Keep trying! Once you've come up with your hypothesis, read on to find out our answer.

4. Start by smearing some water around the mouth of the bottle. The water acts as a lubricant.

5. The next few actions require some good teamwork. Set the glass bottle in front of you on a table. Make sure the water balloon and strip of paper are close by. The adult member of the team is responsible for lighting the strip of paper on fire and quickly pushing it into the bottle. As soon as the burning paper goes into the bottle, the second team member covers the mouth of the bottle with the water balloon.

6. The balloon will immediately start to wiggle around on the top of the bottle, the fire in the bottle will go out, and some invisible force will literally "push" the balloon into the bottle. Amazing!

Now it's on to the next challenge. Can you get the balloon back out of the bottle? Use what you've learned about air and air pressure to come up with a way to get the balloon back out. Here's a hint—try sneaking a straw alongside the balloon when you pull it out. If the outside air can get inside the bottle, the water balloon will come out.

Classic Egg in the Bottle

Now that you've mastered the technique, repeat the six steps (pages 77–78), substituting a hard-boiled egg for the water balloon. The trick here is to find an egg that is just slightly bigger than the mouth of the bottle. The other little secret is to grease the mouth of the bottle with vegetable oil so that the egg slides right in. If you're using the same bottle, make sure you rinse it out with water. This step cleans out the old burnt paper and helps circulate more oxygen into the bottle so the paper will burn. Have an adult light the strip of paper on fire. Carefully push the burning strip of paper into the bottle, quickly cover the mouth of the bottle with the egg, and watch what happens next.

Want to get the egg back out so you can do it again? Try this, if you dare . . . put your mouth over the mouth of the bottle and forcefully blow air into the bottle.

The egg should pop back out of the bottle right into your mouth! Can it get any cooler than that?

Egg in the Bottle—Upside Down Twist

All you need for this variation is a hard-boiled egg, a glass bottle, several birthday candles, and a match. Carefully hold the wider end of the egg in one hand and slowly push two birthday candles into the narrow end of the egg. Light the candles, turn the bottle upside down, and slowly move it into position an inch above the flaming candles. Allow the flames to heat up the air inside the bottle for just a few seconds and then place the bottle down over the candles. The candles will go out and with a "Pop!" the egg will squeeze up into the bottle!

WHAT'S GOING ON HERE?

The burning piece of paper or birthday candles heat the molecules of air in the bottle and cause the molecules to move far away from each other. Some of the heated molecules actually escape out past the egg that is resting on the mouth of the bottle (that's why the egg or water balloon wiggles on top of the bottle). When the flame goes out, the molecules of air in the bottle cool down and move closer together, making room for new air molecules. This is what scientists refer to as a partial vacuum. Normally, the air outside the bottle would come rushing in to fill the bottle. However, that darn egg is in the way! The pressure of the air molecules outside the bottle is so great that it literally "pushes" the egg into the bottle.

REAL-WORLD APPLICATION

When you fly in an airplane or drive high up into the mountains, you've probably noticed that your ears sometimes need to "pop." This "popping" is caused by the same change in air pressure that "pops" the egg into and out of the bottle. Air pressure decreases as altitude increases, so as you go higher, the air pressure decreases, causing the air trapped in your inner ear to push your eardrums outward. Your body tries to regain equilibrium or balance by allowing some of the air in your inner ear to escape through the Eustachian tubes. When the tubes open, the pressure releases and you feel the "pop."

On the way back down to a lower altitude the air pressure increases. The extra pressure from the outside of the ear pushes the eardrums inward. Air moves in through the Eustachian tubes, the ears "pop," and balance is restored. Many people don't wait for this to happen on its own because the pressure imbalance can be uncomfortable. Instead, they just plug their noses, close their mouths, and pretend like they're blowing their noses. Because the air from their lungs has nowhere to go, it is forced into the inner ear through the Eustachian tubes, causing their ears to "pop."

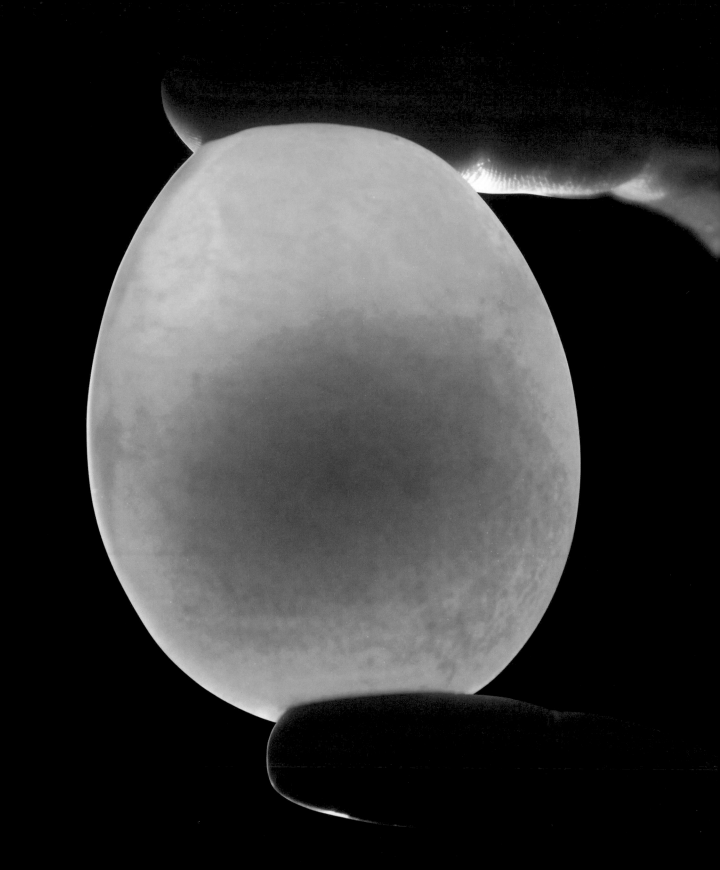

NAKED EGGS

This experiment answers the age-old question, "Which came first, the rubber egg or the rubber chicken?" It's easy to make a rubber, or "naked," egg if you understand the chemistry of removing the hard eggshell. What you're left with is a totally embarrassed naked egg and a cool piece of science.

LET'S TRY IT!

1. Place the egg in a tall glass or jar and cover the egg with vinegar.
2. Look closely at the egg. Do you see any bubbles forming on the shell?
3. Leave the egg in the vinegar for a full 24 hours.
4. Change the vinegar on the second day. Carefully pour the old vinegar down the drain and cover the egg with fresh vinegar. Place the glass with the vinegar and egg in a safe place for a week—that's right, 7 days! Don't disturb the egg but pay close attention to the bubbles forming on the surface of the shell (or what's left of it).
5. One week later, pour off the vinegar and carefully rinse the egg with water. The egg looks translucent because the outside shell is gone! The only thing that remains is the delicate membrane of the egg. You've successfully made an egg without a shell. Okay, you didn't really make the egg (the chicken made the egg), you just stripped away the chemical that gives the egg its strength.

TAKE IT FURTHER

Do organic or free-range eggs have an eggshell that is stronger or weaker than generic eggs? Conduct your own test on several different kinds of eggs all at once to observe any differences in the time required for the vinegar to dissolve the shells.

Try using concentrated vinegar instead of traditional vinegar. Concentrated vinegar is about four times the strength of traditional household vinegar. If you really want to cut down on the time it takes for the eggshell to disappear . . . and you're a chemistry teacher . . . try using 1 molar hydrochloric acid. Be careful—this is really strong stuff!

Bouncing Eggs

Here's another idea. Put an egg in a separate glass. Cover the egg with vinegar. Allow the egg to sit in the vinegar for 24 hours. After 24 hours, pour out the vinegar and take the egg out of the glass. Drop the egg into the sink from a height of 3 inches. What happens? Continue dropping the egg from different heights (all drops should be done over the sink). What is the greatest height that you can drop the egg from before the egg goes splat? Can you measure the height of the bounces?

Secret Message Eggs

You'll totally freak someone out by making a message appear on an egg. Start by boiling an egg in a saucepan on the stove for 10 minutes. Remove the egg from the pan and let it cool. Use a crayon or small candle to write on the eggshell. Write anything you want—your name, a design, a symbol—just like you were going to make wax designs on an Easter egg. Then place the egg in a glass filled with vinegar. Bubbles will begin to form on the surface of the egg. When the bubbling stops, pour out the vinegar and cover the egg with fresh vinegar. When the second round of bubbling has stopped, remove the egg from the glass of vinegar and rinse it off with cool water. Rub your fingers over the surface of the egg. What do you feel? What do you see? The eggshell is gone, but you should be able to decipher what you wrote or drew on the eggshell. Wax does not react with acid (vinegar), so underneath your wax design, the eggshell remains intact.

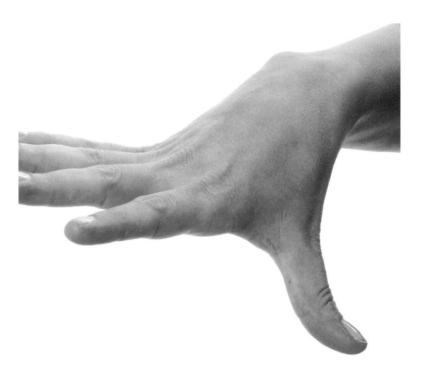

WHAT'S GOING ON HERE?

Let's start with the bubbles you saw forming on the shell. The bubbles are carbon dioxide

CO_2). Vinegar is an acid called acetic acid (CH_3COOH), and white vinegar from the grocery store is usually about 4% acetic acid and 96% water. Eggshells are made up of calcium carbonate ($CaCO_3$). The acetic acid in the vinegar reacts with the calcium carbonate in the eggshell to make calcium acetate plus water and carbon dioxide bubbles that you see on the surface of the shell.

The chemical reaction looks like this . . .

$$2\ CH_3COOH + CaCO_3\ =\ Ca(CH_3COO)_2 + H_2O + CO_2$$
Acetic acid + Calcium carbonate = Calcium acetate + Water + Carbon dioxide

The egg looks translucent when you shine a flashlight through it because the hard outside shell is gone. The only part that remains is the thin membrane called a semipermeable membrane.

You might have noticed that the egg got a little bigger after soaking in the vinegar. Here's what happened . . . Some of the water in the vinegar solution (remember that household vinegar is 96% water) traveled through the egg's membrane in an effort to equalize the concentration of water on both sides of the membrane. This flow of water through a semipermeable membrane is called **osmosis**.

If you take your naked egg and place it in a glass filled with corn syrup, the egg will shrivel. Since corn syrup has a lower concentration of water than an egg does, the water in the egg moves through the membrane and into the corn syrup to equalize the water concentration levels on both sides.

HOW TO MAKE A **FOLDING EGG**

Just imagine the look on your friends' faces when you show them an egg and then proceed to fold it in half several times until it forms a small white ball! Wait, it gets better. Just bounce the "folded egg" between your hands and the egg reappears!

LET'S TRY IT!

1. The first step is the trickiest and requires a little practice. You'll need to blow out the inside of the egg without causing too much visible damage. With the help of an adult, use a sharp pin, a thumbtack, or the tip of a sharp knife to poke a small hole in both ends of the egg. You can also use a small drill to make the holes in the egg. The hole should be about ⅛ inch in diameter. Don't be frustrated if you crack a few eggs before you get the hang of it.

2. The next step is to scramble the inside of the egg in order to break the yellow yolk. The best way to break the yolk is to poke a toothpick or something similar (like a plastic coffee stirrer) through the hole and to poke around carefully inside the egg.

3. Once the yolk is broken and the egg is "scrambled," it's time to blow all of the liquid out of the egg. One method is to clean off one end of the egg with a moist towelette, cover the hole with your mouth, and blow the egg liquid out of the other hole. Of course, it's best to hold the egg over the sink as you're doing this. People who are concerned about using their mouths may choose not to try the activity.

4. Place the hollow egg in a tall glass or jar and cover the egg with vinegar. You want the egg to be completely submerged in the vinegar, which means that you may need to place something like a large spoon on top of the egg to push it down. You can also try filling the inside of the egg with vinegar to weigh it down.

5. Leave the egg in the vinegar for a full 24 hours.

6. Change the vinegar on the second day. Carefully pour the old vinegar down the drain and cover the egg with fresh vinegar.

Place the glass with the vinegar and egg in a safe place for up to 10 days or until all of the shell has dissolved. Some eggshells will take longer to dissolve than others because every egg is unique. For the first few days, bubbles of carbon dioxide (CO_2) will form on the shell. The vinegar is dissolving the calcium carbonate in the shell and producing bubbles of CO_2 at the same time. When the bubbles stop forming, it's a good indication that the eggshell has completely dissolved.

7. Once the bubbles have stopped forming (again, this could take up to 10 days so be patient!) pour off the vinegar and carefully rinse the egg membrane with water.

8. Carefully squeeze out all of the water from inside the egg membrane. Gently blow a little air into one end of the egg and the egg will puff up. Hey, it looks like a real egg! Slowly squeeze the egg in your hand and it will look like you crushed the egg. Carefully toss and bounce the "folded egg" in your hands to magically restore its shape.

There's both an art and science to making a good folding egg. It just takes practice and patience until you get a membrane that really works. The more you toss the membrane between your hands, the more the water evaporates from the membrane, air gets pushed into the membrane, and it begins to take on the shape of a real egg.

TAKE IT FURTHER

Dust the almost dry egg membrane with some baby powder (sometimes called talcum powder). Try to get some of the powder inside the egg as well. The powder helps keep the egg membrane from drying out and cracking, and it makes the egg look even more real.

WHAT'S GOING ON HERE?

As you already know from the Naked Egg experiment, the acetic acid in the vinegar breaks down the calcium carbonate in the eggshell, and the bubbles that form on the surface of the egg are CO_2. Eventually the hard shell of the egg disappears entirely and all that remains is the egg membrane. Because you have already blown out the contents of the egg, the membrane is just full of air. You can fold it up and the air will sneak out the tiny hole in the membrane that you used to blow the yolk out of the egg. The membrane will compress down into practically nothing. As you gently toss around and bounce the "folded egg" on your hand, the air will reenter the membrane, expanding back into its original shape and volume.

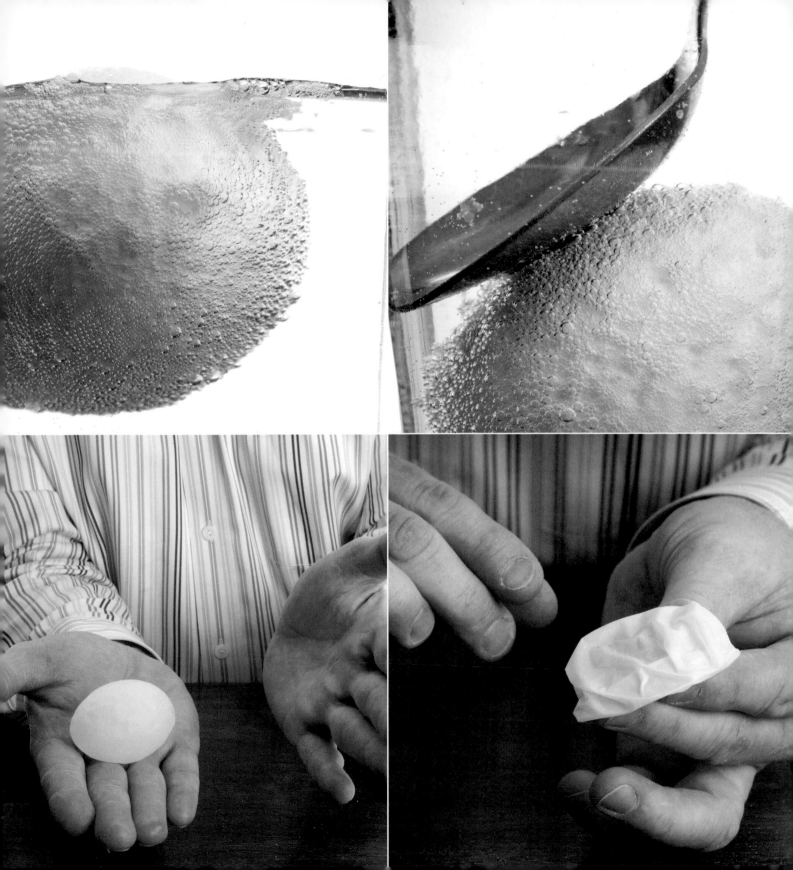

DRY ICE

There's just something magical about a piece of dry ice. Drop a chunk of it into a glass of water and watch as white smoke erupts from the bubbling concoction. Once you witness that, you'll be hooked.

Dry ice is not frozen water—it's frozen carbon dioxide (CO_2). Unlike most solids, dry ice does not melt into a liquid, but instead changes directly into a gas. This process is called **sublimation**. The temperature of dry ice is -109.3°F or -78.5°C. Dry ice does not last very long—the experts tell us that it will sublimate at a rate of 5 to 10 pounds every 24 hours in a typical Styrofoam chest. So it's important to purchase the dry ice as close as possible to the time you need it. If you're planning to perform a number of dry ice demonstrations, purchase 5 to 10 pounds.

Just remember these safety rules when using dry ice . . .

- Only use dry ice with adult supervision.

- Dry ice must be handled using gloves or tongs because it will cause severe burns if it comes in contact with your skin.

- Never put dry ice in your mouth.

- Never store dry ice in an airtight container. The gas will build up and the container will explode. Make sure your container is ventilated. The best place to store dry ice is in a Styrofoam chest with a loose-fitting lid to allow the CO_2 to escape.

- Do not store dry ice in your freezer. It will cause your freezer to become too cold and your freezer may shut off. On the other hand, if you lose power for an extended period of time, dry ice is the best way to keep things cold.

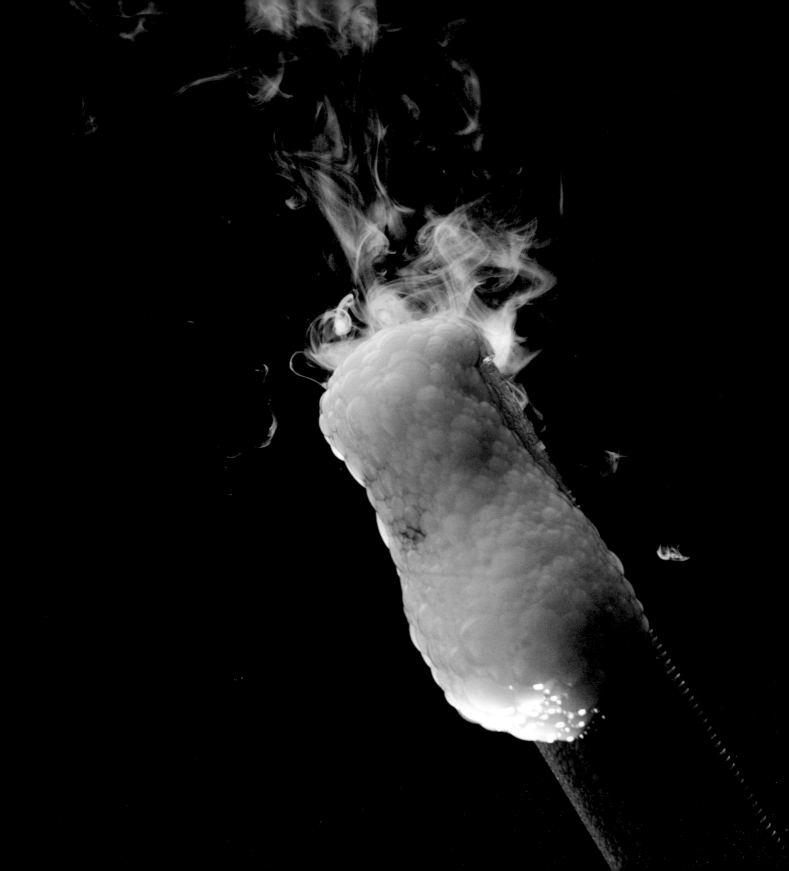

SUBZERO SCIENCE—
COOL SCIENCE WITH DRY ICE

Halloween is the perfect time for smoking, bubbling, eye-catching potions. But you don't have to wait for Halloween to have fun and learn about the states of matter using dry ice. Now that you've read the safety rules, try these awesome dry ice experiments and prepare to be amazed!

WHAT YOU NEED

Dry ice

Heavy duty gloves or tongs

Water

Liquid dish soap
(Dawn works well)

Drinking glass

Plastic cylinder (optional)

Food coloring

Glowing light sticks
(optional)

Small fish aquarium

Bubble wand and
bubble solution

Disappearing Ice

Here's a quick experiment to help friends better understand why it's called dry ice. Use tongs to place a regular ice cube on one plate and a similar size piece of dry ice on a second plate. Keep both plates out of the reach of children. If you ask your friends to predict what will happen to the ice during the next few hours, most will likely say that the pieces of ice will turn into puddles of water.

Allow everyone to view the plates after 1 hour and they'll discover the difference between real ice and dry ice. There should be a puddle of water on the plate where the real ice was, but the dry ice plate will be "dry." Where did the dry ice go? Dry ice is not made from water, it's frozen carbon dioxide (CO_2). The dry ice turned into invisible CO_2 that disappeared into the air. Magic? No, it's science.

Burping, Bubbling, Smoking Water

Use the tongs or gloves to place a piece of dry ice in a glass of warm water. Immediately, the dry ice will begin to turn into CO_2 and water vapor, forming a really cool cloud. This cloud is perfectly safe for you to touch and feel as long as you are careful not to reach into the water and accidentally touch the dry ice.

To create the best effect, be sure to use warm water. Over time, the dry ice will make the water cold and the "smoking" will slow down. Replace the cold water with warm water and you're back in business. While the term "smoke" is often used to describe what you see, it's technically a cloud of water vapor fog.

Smoking Bubbles

Who would have guessed that you could have this much fun with soapy water and a chunk of dry ice? Fill a tall glass or plastic cylinder with warm water and add a squirt of liquid dish soap like Dawn or Joy. Use gloves or tongs to place a piece of dry ice into the soapy water.

Instead of the dry ice just bubbling in the water to make a cloud, the soap in the water traps the CO_2 and water vapor in the form of a bubble. The bubbles climb out of the cylinder of warm, soapy water and explode with a burst of misty fog as they crawl over the edge.

Add some food coloring to the water to make the demonstration more colorful. If you want to give the bursting suds an eerie glow, drop a glowing light stick into the water along with the dry ice. The light stick will give the bursting bubbles an eerie look, perfect for any Halloween party.

Make a Bubbling Beverage

The next time you have a craving for a sparkling beverage, make your own batch using what you know about dry ice. Fill a bowl or pitcher with apple juice and use tongs to add a few large pieces of dry ice. While the mixture is bubbling and burping, the apple juice is being carbonated by the dry ice. CO_2 mixes with the juice to make a "sparkling" drink. Your local hobby or craft store is sure to have a spooky-looking Halloween cauldron that would hold a large batch of apple juice and dry ice. Wait until the dry ice is completely gone before serving the apple juice. It's a spooky carbonated drink.

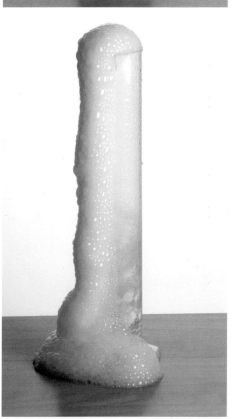

Floating Bubble

You'll notice that when you add dry ice to water, the cloud of CO_2 and water does not float up into the air, but instead falls toward the ground. Why? This cloud-like mixture of CO_2 and water is heavier than the surrounding air. You'll use this little piece of science trivia to perform the amazing "Floating Bubble" trick.

A small fish aquarium works well for this activity. Fill the bottom of the aquarium with about an inch of warm water (take the fish out first!). Use gloves or tongs to add a few pieces of dry ice. Of course, the dry ice will begin to smoke as it turns into CO_2 and water vapor.

Using a bubble wand and a bottle of bubble fluid, blow a few bubbles into the aquarium (it's a little difficult so be patient). To everyone's amazement, a few bubbles will appear to float in midair inside the aquarium. The bubble is really just floating on a cushion of invisible CO_2. Of course, the spooky Halloween story is up to you . . . but I think I heard that the aquarium is the home of a ghost who has been known to play with soap bubbles!

WHAT'S GOING ON HERE?

By doing the activities described in this section, you've learned that dry ice is frozen CO_2. Under normal atmospheric conditions, CO_2 is a gas. Only about 0.035% of our atmosphere is made up of CO_2. Most of the air we breathe is nitrogen (79%) and about 20% is oxygen. CO_2, along with a handful of other gases, make up the remaining 1% of the air we breathe.

Did you know that dry ice is often mixed with regular ice to save shipping weight and extend the cooling energy of water ice? Dry ice gives more than twice the cooling energy per pound of weight and three times the cooling energy per volume than regular water ice.

How is dry ice made? The first step in making dry ice is to compress CO_2 until it liquefies, at the same time removing the excess heat. The CO_2 will liquefy at a pressure of approximately 870 pounds per square inch at room temperature. Once liquid CO_2 is formed, the CO_2 is sent through an expansion valve and enters a pressure chamber. The pressure change causes the liquid to flash into a solid and causes the temperature to drop quickly. About 46% of the gas will freeze into "dry ice snow." The rest of the CO_2, 54%, is released into the atmosphere or is recovered to be used again. The dry ice snow is then collected in a chamber where it is compressed into block, pellet, or rice-size pieces using hydraulics.

DRY ICE **CRYSTAL BALL BUBBLE**

Create a soap film on the rim of a bucket and, with one other simple ingredient,
you will have made the world's coolest crystal ball.

LET'S TRY IT!

1. Select a bowl that has a smooth rim and is smaller than 12 inches in diameter.
2. Mix 2 tablespoons (30 mL) of Dawn liquid dish soap with 1 tablespoon (15 mL) of water in a plastic cup.
3. Cut a strip of cloth about 1 inch (2.5 cm) wide and 18 inches (46 cm) long. Soak the cloth in the soapy solution, making sure that the cloth is completely submerged.
4. Fill the bowl half full with warm water. Have gloves ready to transfer the dry ice to the bowl.
5. Place two or three pieces of dry ice into the water so that a good amount of fog is produced.
6. Remove the strip of cloth from the soap solution and run your fingers down the cloth to remove the excess soap. Stretch the cloth between your hands and slowly pull the soapy cloth across the rim of the bowl. The goal is to create a soap film that stretches across the entire bowl. It also helps to dip your fingers in some water and wet the rim of the bowl before you start. Getting the soap film to stretch across the rim of the bowl can take a little practice until you get the technique mastered. If all else fails, try cutting a new strip of cloth from a different type of fabric (an old T-shirt works well), or change the soap solution by adding some more water.

TAKE IT FURTHER

If you accidentally get soap in the bowl of water, you'll notice that zillions of bubbles filled with fog will start to emerge from the bowl. This, too, produces a great effect. Place a waterproof flashlight in the bowl along with the dry ice so that the light shines up through the fog. Draw the cloth across the rim to create the soap film lid and, if you are inside, turn off the lights. The crystal bubbles will emit an eerie glow and you'll be able to see the fog churning inside the transparent bubble walls. When the giant bubble bursts, the cloud falls to the floor, followed by an outburst of ooohs and ahhhs from your audience!

WHAT'S GOING ON HERE?

When you drop a piece of dry ice in a bowl of water, the gas that you see is a combination of carbon dioxide and water vapor. So, the gas that you see is actually a cloud of tiny water droplets. The thin layer of soap film stretched across the rim of the bowl traps the expanding cloud to create a giant bubble. When the water gets colder than 50°F, the dry ice stops making fog, but continues to sublimate and bubble. Just replace the cold water with warm water and you're back in business.

GOOEY WONDERS

CORNSTARCH SCIENCE—
QUICKSAND GOO

Anyone who has ever watched a classic Western knows about the dangers of quicksand. You know, that gooey stuff that grabs hold of its victim and swallows him alive? So, what is quicksand and how does it really work? In this experiment, you'll use ordinary cornstarch to model the behavior of real quicksand.

WHAT YOU NEED

One box of cornstarch
(16 ounces)

Large mixing bowl

Cookie sheet, square cake
pan, or something similar

Pitcher of water

Spoon

Gallon-size zipper-lock bag

Newspaper or a plastic drip
cloth to cover the floor

Water

Food coloring (optional)

LET'S TRY IT!

1. Pour approximately one-quarter of the box (about 4 ounces) of cornstarch into the mixing bowl and slowly add about ½ cup of water. Stir. Sometimes it is easier (and more fun!) to mix the cornstarch and water with your bare hands.

2. Continue adding cornstarch and water in small amounts until you get a mixture that has the consistency of honey. It may take a little work to get the consistency just right, but you will eventually end up mixing one box of cornstarch with roughly 1 to 2 cups of water. As a general rule of thumb, you're looking for a mixture of roughly ten parts cornstarch to one part water. Notice that the mixture gets thicker, or more viscous, as you add more cornstarch.

3. Sink your hand into the bowl of "quicksand" and notice its unusual consistency. Compare what it feels like to move your hand around slowly and then very quickly. You can't move your hand around very fast. In fact, the faster you thrash around, the more like a solid the gooey stuff becomes. Sink your entire hand into the goo and try to grab the fluid and pull it up. That's the sensation of sinking in quicksand!

4. Drop a plastic toy animal into the cornstarch mixture and then try to get it out. It's pretty tough even for an experienced quicksand mixologist.

TAKE IT FURTHER

Pour the mixture onto the cookie sheet or cake pan. Notice its unusual consistency when you are pouring it onto the pan. Stir it around with your finger, first slowly and then as fast as you can. Skim your finger across the top of the glop. What do you notice?

Try to roll the fluid between your palms to make a ball. You can even hold your hand flat over the top of the pan and slap the liquid glop as hard as you can. Most people will run for cover as you get ready to slap the liquid, fearing that it will splash everywhere.

According to theory, the mixture should stay in the pan. Yeah, right! If your cornstarch-water mixture inadvertently splatters everywhere, you will know to add more cornstarch. When you are finished, pour the glop into a large zipper-lock bag for later use.

WHAT'S GOING ON HERE?

The cornstarch and water mixture acts like a solid sometimes and a liquid at other times. This concoction is an example of a **suspension** (a mixture of two substances), one of which is finely divided and dispersed in the other. In the case of the cornstarch quicksand, it's a solid dispersed in a liquid.

When you slap the cornstarch quicksand, you force the long starch molecules closer together. The impact of this force traps the water between the starch chains to form a semirigid structure. When the pressure is released, the cornstarch flows again.

All fluids have a property known as **viscosity**—the measurable thickness or resistance to flow in a fluid. Honey and ketchup are liquids that have a high resistance to flow, or a high viscosity. Water has a low viscosity. Sir Isaac Newton said that viscosity is a function of temperature. So, if you heat honey, the viscosity is less than that of cold honey. The cornstarch-water mixture, and real quicksand, are both examples of **non-Newtonian** fluids because their viscosity changes when stress or a force is applied, not when heat is applied.

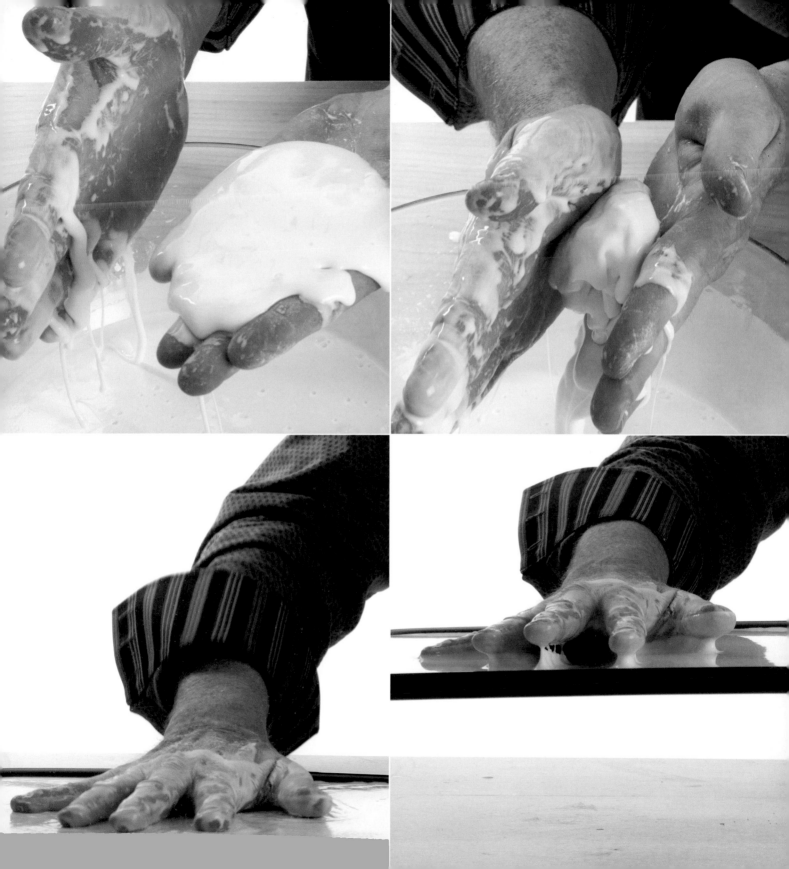

What is quicksand, really? Quicksand is nothing more than a soupy mixture of sand and water, where the sand is literally floating on water. Quicksand is just solid ground that has been liquefied by too much water, and the term "quick" refers to how easily the sand shifts in this solid-liquid state. Quicksand is created when water floods or saturates an area of loose sand and the sand begins to move around. When the water in the sandy soil cannot escape, it creates a liquid-like soil that can no longer support any weight. If an excessive amount of water flows through the sand, it forces the sand particles apart. This separation of particles causes the ground to loosen, and any weight placed on the sand will begin to sink through it.

The quicksand phenomenon can be caused by flowing underground water rising to the surface or even an earthquake that agitates the sand. You are likely to find quicksand around riverbanks, lake shorelines, marshes, beaches, near underground springs or any place where an uprising of water oversaturates and agitates the sand.

The next time you are standing barefoot on the beach, think about the properties of quicksand. Normally, the grains of wet sand are compressed together tightly and this firm ground easily supports your weight. The friction between grains of wet sand is strong enough to make it easy to build sand castles. However, when the sand on the beach is flooded with an excess amount of water, the agitated sand particles begin to move, separate, and quickly wash away right out from under your feet.

Steve on *The Ellen DeGeneres Show*

In an effort to live out our mission ("Make It Big, Do It Right, Give It Class!"), we took the Quicksand Goo experiment to a whole new level on national television in February 2008. When we originally discussed the idea of mixing up a batch of cornstarch and water with the producers of *The Ellen DeGeneres Show,* there was little interest—because they thought it was just a little tabletop experiment. When we shared the bigger idea of mixing up a batch of cornstarch goo big enough to fill a giant tub and inviting Ellen to walk across the surface without sinking, the producers gave us the green light.

The challenge was to turn this kitchen chemistry activity into a large-scale demonstration. Rather than buying one box of cornstarch, we were going to need 2,500 boxes of cornstarch. Anyone who has ever played with this cornstarch and water mixture knows that the bigger the batch gets, the harder it is to mix. If Ellen or someone else was going to walk on water, my team and I would have to find a way to mix up a huge batch. That's when my team of mixologists pitched the idea to the Bonanza Concrete company in Burbank, California. If 1 pound of cornstarch in a mixing bowl is fun, 2,500 pounds in a cement truck should be a blast.

Our goal was to mix up a batch large enough to fill a giant bathtub and then to attempt a quick run across the surface. The weight of someone's body should cause the liquid to turn into a solid for a split second, allowing the person to literally run across the surface. Well, that's the theory anyway.

Let's start with the giant bathtub. The props guys on *The Ellen DeGeneres Show* (a very creative and talented team of people who can pull off anything) constructed a container 7 feet long, 3 feet wide, and 1½ feet deep that held about 240 gallons of water. After mixing up several test batches, we discovered that the mixing ratio of ten parts cornstarch to one part water worked beautifully. Given the size of the tub, we would need roughly 2,400 pounds of cornstarch to mix with the 240 gallons of water.

Maybe it's best to use pictures to explain the next part of the story . . .

Since our original stunt aired in February 2008, the Bonanza Concrete company has mixed up five more batches of cornstarch goo for game shows, educational programs, and even a competition on a reality TV show. Back to our motto . . . "Make It Big, Do It Right, Give It Class." Classy? Maybe not, but it sure made an impression!

Get 2,500 pounds of cornstarch . . . and a forklift to move all those bags.

Assemble a team of cornstarch mixologists who can share in the responsibility if something goes wrong (this is a very important strategic move on my part).

Convince the owners of Bonanza Concrete that it's a good idea for us to mix up a batch of experimental cornstarch goo in their truck.

Drive the mix to *The Ellen DeGeneres Show* studios in Burbank, California.

Quickly discover that the cornstarch mixture needs constant mixing until the show begins and the props people forklift the container into the studio. If all else fails, at least we've come up with a new workout routine.

Off-load the cornstarch and water mixture into the giant bathtub in the parking lot of the studio and cross your fingers that the mix is still good.

Ellen DeGeneres invites an audience member to dance across the giant tub of cornstarch goo . . . and the audience goes wild!

GLACIER **GAK**

If you're old enough to remember Silly Putty, you may remember what it felt like to roll it up into a ball in your hand. The fascinating thing about Silly Putty is that you can put a ball of it down on the table and, eventually, it spreads itself out. How does it do that? Here's a simple way to make your own version of this popular toy called GAK using household materials. You'll use the recipe to make a white and a blue batch of GAK, give them a little twist, and use the concoction to learn more about the movements of real glaciers.

WHAT YOU NEED

Elmer's Glue-All
(two 8-ounce bottles
to make two batches)

Borax (a powdered soap
found in the grocery store)

Empty plastic soda
bottle with cap

Large mixing bowl

Plastic cup (8-ounce size)

Spoon

Cookie sheet or plastic tray

Measuring cup

Food coloring

Water

Paper towel

Zipper-lock bag
(don't you want to keep
it when you're done?)

LET'S TRY IT!

1. You'll make two separate batches of GAK—one will be white and the other will be blue. You could purchase one large container of Elmer's Glue-All and follow the recipe or simply use two 8-ounce bottles. The measurements do not have to be exact, but it's a good idea to start with the suggestions below for the first batch.

2. This recipe is based on using a brand-new, 8-ounce (240 mL) bottle of Elmer's Glue-All. Empty the entire bottle of glue into a mixing bowl. Fill the empty bottle half full with warm water and shake (okay, put the lid on first and then shake). Pour the glue-water mixture into the mixing bowl and use the spoon to mix well.

3. Measure ½ cup (120 mL) of warm water into the plastic cup and add a heaping teaspoon of Borax powder to the water. Stir the solution. Don't worry if all of the powder dissolves.

4. While stirring the glue in the mixing bowl, slowly add a little of the Borax solution you just made. Immediately you'll feel the long strands of molecules start to connect. It's time to abandon the spoon and use your hands to do the serious mixing. Keep adding the Borax solution to the glue mixture (don't stop mixing) until the GAK has a putty-like consistency. You should be able to roll it on the table like dough, but if you let it rest for a couple of minutes, the GAK will spread itself out. Set this batch of white GAK off to the side while you mix up the next batch.

5. Repeat the four preceding steps to make your second batch of GAK, but this time add about ten drops of blue food coloring during Step 2. As you're adding the Borax solution, try to keep the consistency of this batch the same as that of the one you previously made.

6. Here comes the fun part of combining the blue and white batches of GAK. There's no right or wrong way to do this—just twist and fold the large pieces until you get a cool swirl of blue and white GAK.

7. Lay the GAK out on a cookie sheet or plastic tray. Use a few books to prop up one end of the cookie sheet. This will allow the GAK to begin to flow slowly downhill. The mixture of blue and white GAK creates some amazing patterns as the mixture flows downward, simulating the slow movement of a glacier.

8. At any point you can pick up the GAK, reshape it into a big blob, set it at the top of the cookie sheet, and restart the flow. Try placing various sizes of rocks on the cookie sheet as obstacles for the GAK to flow around. Depending on the size of the rock, the movement of the Glacier GAK might be powerful enough to push the rock down the hill. You can read more about the science of glaciers in the "Real-World Application" section on page 115.

9. When you're finished playing with your GAK, seal it up in a zipper-lock bag for safekeeping.

TAKE IT FURTHER

Adjust the amounts of water, Borax, and glue to see how the consistency of the GAK changes. Remember to change only one ingredient at a time to best determine the effect each ingredient has on the overall consistency of the putty.

WHAT'S GOING ON HERE?

The mixture of Elmer's Glue-All with Borax and water produces a putty-like material called a **polymer**. In simplest terms, a polymer is a long chain of molecules. You can use the example of cooking spaghetti to better understand why the GAK polymer behaves the way it does. When a pile of freshly cooked spaghetti comes out of the hot water and into the bowl, the strands flow like a liquid from the pan to the bowl. This is because the spaghetti strands are slippery and slide over one another. When the water drains off of the pasta, the strands start to stick together and the spaghetti takes on a rubbery texture. Wait a little while longer for all of the water to evaporate, and the pile of spaghetti turns into a solid mass. Drop it on the floor and watch it bounce.

Many natural and synthetic polymers behave in a similar manner. Polymers are made out of long strands of molecules like spaghetti. If the long molecules slide past each other easily, then the substance acts like a liquid because the molecules flow. If the molecules stick together at a few places along the strand, then the substance behaves like a rubbery solid called an **elastomer**. Borax is the compound that is responsible for hooking the glue's molecules together to form the putty-like material. There are several different methods for making this putty-like material. Some recipes call for liquid starch instead of Borax soap. Either way, when you make this homemade Silly Putty you are learning about some of the properties of polymers.

REAL-WORLD APPLICATION

For hundreds of thousands of years, the movement of glaciers has shaped land through erosion and deposition, creating landforms such as U-shaped valleys, moraines, and kettle lakes, to name a few. Motion and change define a glacier's life, and glaciers grow and shrink in response to a changing climate. Currently, glacial retreat is implicated in the Earth's changing climate patterns and may have a great impact on sea levels and weather cycles.

The unique, slow moving properties of the GAK simulate the movement of a glacier. At a molecular level, ice is comprised of stacked layers of molecules with relatively weak bonds between the layers. This is similar to the makeup of our GAK molecules. Ice can stretch or break depending on the amount of pressure applied. If there is a lot of pressure or a high strain rate, ice will crack or break (causing crevasses in glaciers). When the pressure is lower or the strain rate is small and constant, ice can bend or stretch. The steady pressure from the bulk of the ice mass and the pull of gravity cause the glacier to flow slowly (so slowly you can't see it) downhill, bending like a river of ice.

Scientists study glaciers because by their movement, glaciers mark change over an incredible period of time. By monitoring glaciers around the world over time, researchers construct valuable records of glacial activity and their response to climate variation.

THE **BABY DIAPER** SECRET

If you've ever changed a diaper and noticed what looked like tiny crystals on the baby's skin, you've uncovered the secret of superabsorbent, disposable diapers. Those tiny crystals actually come from the lining of the diaper and are made out of a safe, nontoxic polymer that absorbs moisture away from the baby's skin. This amazing polymer changed the way parents care for their babies, and scientists continue to find new uses for these superabsorbent polymers.

WHAT YOU NEED

Disposable diapers
(several brands)

Plastic cup (8-ounce size)

Zipper-lock bag

Scissors

Newspaper

Water

LET'S TRY IT!

1. Place a new (unused is your first choice) diaper on the piece of newspaper. Carefully cut through the inside lining and remove all the cotton-like material. Put all the stuffing material into a clean, zipper-lock bag.

2. Scoop up any of the polymer powder that may have spilled onto the paper and pour it into the bag with the stuffing. Blow a little air into the bag to make it puff up like a pillow, then seal the bag.

3. Shake the bag for a few minutes to remove the powdery polymer from the stuffing. Notice how much (or how little) powder falls to the bottom of the bag.

4. Carefully remove the stuffing from the bag and check out the dry polymer you just extracted from the diaper.

5. Pour the polymer into a plastic cup and fill the cup with about 4 ounces (120 mL) of water. Mix it with your finger until the mixture begins to thicken.

6. Observe the gel that the polymer and water create. Turn the cup upside down and see how it has solidified. Now you know the super, moisture-absorbing secret hiding in the lining of a baby diaper. You just discovered something that has both a cool and a yuck factor!

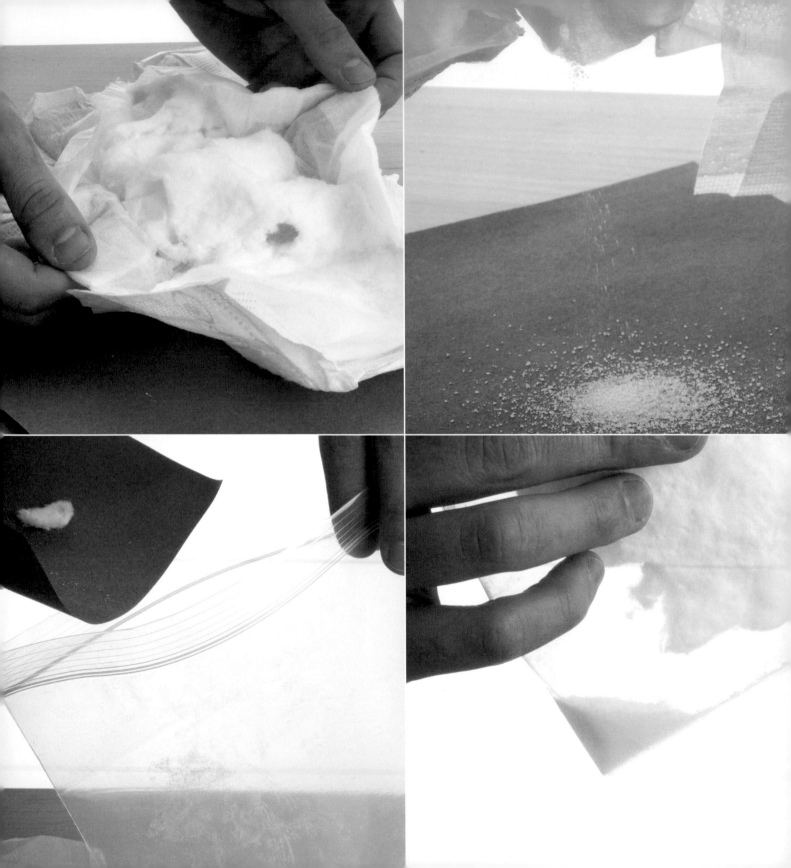

Put the pieces of gel back into the cup and smush them down with your fingers. Add a teaspoon of salt, stir it with a spoon, and watch what happens. Salt messes up the gel's water-holding abilities. When you're finished, pour the saltwater goo down the drain.

Grab a new diaper and slowly pour about ¼ cup of warm tap water into the center. Hold the diaper over a large pan or sink and continue to add water, a little at a time, until it will hold no more. Keep track of how much water the diaper can absorb before it reaches its limit.

WHAT'S GOING ON HERE?

The secret, water-absorbing chemical in a diaper is a superabsorbent polymer called sodium polyacrylate. A **polymer** is simply a long chain of repeating molecules. If the prefix "poly" means many, then a polymer is a long chain of molecules made up of many smaller units, called **monomers**, which are joined together. Some polymers are made up of millions of monomers.

Superabsorbent polymers expand tremendously when they come in contact with water because water is drawn into and held by the molecules of the polymer. They act like giant sponges. Some can soak up as much as 800 times their weight in water. Just imagine how much water a giant diaper could hold (then again, don't . . . that's gross).

The cotton-like fibers you removed from the diaper help to spread out both the polymer and the, uh, "water" so that the baby doesn't have to sit on a mushy lump of water-filled gel. This explanation is getting grosser and grosser! It's easy to see that even a little bit of polymer powder will hold a huge quantity of water, but it does have its limits. At some point, the baby will certainly let you know that the gel is full and it's time for new undies!

In spite of their usefulness, these diapers can be a problem. If you've ever observed a baby in diapers splashing in a wading pool, you know that even one diaper can absorb lots and lots of water. Most public pools won't allow them to be worn in the water because huge globs of gooey gel can leak out and make a mess of the filter system. Also, some folks used to throw them away in toilets—not a good idea unless you're a plumber. For the most part, however, these diapers are a great invention and make for dry, happy babies.

REAL-WORLD APPLICATION

Today, superabsorbent polymers are widely used in such applications as forestry, gardening, and landscaping as a means of conserving water. Imagine using a substance that could store water in the soil and then release it as the plants' roots needed it. While we may consider water-absorbing polymers to be a modern convenience, the impact that such technology is having on parts of the world that are plagued by drought is remarkable.

DON'T TRY THIS
AT HOME . . .

TRY IT AT A
FRIEND'S HOME!

FLOATING WATER

Is it really possible to fill a glass with water and turn it upside down without spilling?
This clever science trick is a popular after-dinner science stunt, but make
sure there's a bowl close by to catch your mistakes.

Plastic cup or
drinking glass

Index card or
old playing card

Large bowl or sink
to practice over

LET'S TRY IT!

1. Before you get started, make sure the index card or playing card is large enough to completely cover the mouth of the glass. Fill the glass or plastic cup to the top with water.

2. Cover the cup with an old playing card, making sure that the card completely covers the mouth of the container.

3. Keep your hand on the card and turn the cup upside down. Hold the cup over the bowl just in case you accidentally spill.

4. The final step takes guts. Slowly take your hand away and the card will stay in place . . . and so should the water (keep your fingers crossed).

5. Don't press your luck too far. Put your hand back on the card and return the cup to its upright position.

6. The temptation is just too great, and you know you're going to do it again. Just make sure the card doesn't become completely soaked and accidentally fall apart. This could be a huge surprise for everyone!

TAKE IT FURTHER

Repeat the experiment but this time change the amount of water in the cup. Does it make any difference? What about if you switch the container? Will a wider cup hold the

card better than a narrower cup? Does the temperature of the water have any effect on the water staying inside the cup?

Try the experiment using a paper cup or plastic cup but this time, using a thumbtack, poke a small hole in the bottom of the cup. What do you predict will happen if air is allowed to sneak into the cup?

WHAT'S GOING ON HERE?

The secret is right in front of your nose—it's the air that we breathe. Air molecules in the atmosphere exert pressure on everything. Scientists know that at sea level air molecules in the atmosphere exert almost 15 pounds of pressure (okay, 14.7 pounds if you want to be exact) per square inch of surface area. Your body is used to feeling this kind of air pressure, so you don't notice it.

When you first turn the cup upside down, the pressure of the air inside the cup and the air pressure outside the cup are equal. If you look closely, however, you'll notice that just a little water leaks out between the card and the cup. This happens because the force of gravity naturally pulls down on the water. When some of the water escapes, this causes the volume of air (the space above the water inside the cup) to increase slightly. Even though the amount of air above the water stays the same, the volume occupied by the air is now greater and the air pressure inside the cup decreases. The pressure of the air outside the cup is now greater than the pressure inside the cup and the card stays in place. All of this is possible because the water creates an airtight seal between the rim of the cup and the card.

When the seal is broken (even a *tiny* bit), air enters into the cup, equalizes the pressure, and gravity pushes the water out. Poking a thumbtack-size hole in the cup allows air to seep into the cup from the outside. The pressure of the air molecules both inside and outside the cup stays the same, gravity takes over, the card falls, and the water spills. Watch out for the carpet!

THE **LEAKPROOF** BAG

Who would have ever thought that a plastic bag, some water, and a few pencils would have adults screaming with fear? Learn how to poke holes in a plastic bag filled with water without spilling a drop. Well, that's the theory you're going to test . . . and it's wise to practice your liquid trick over the sink. It's a cool way to learn about the chemistry of polymers.

WHAT YOU NEED

Five sharpened pencils

Zipper-lock plastic bag
(quart-size works well)

Water

A few paper towels

HELPFUL HINTS:

Make sure the tips of the pencils are sharpened to a point. Be careful not to push the pencils all the way through the holes or your "spear-it" experiment will turn into a big "clean-it-up" activity.

LET'S TRY IT!

1. Start by sharpening the pencils. Make sure the tips are sharpened to a point.

2. Fill the bag one-half full with water and then seal the bag closed. Pose this question to your dinner guests: "What would happen if I tried to push one of these pencils through the bag of water? Would the water leak out and make a giant mess?"

3. They'll answer "yes," unless they know the scientific secret, but you're not telling . . . yet!

4. Here comes the scary part. Hold the pencil in one hand and the top of the bag in the other hand. Believe it or not, you can push the pencil right through one side of the bag and halfway out the other side without spilling a drop. The long chains of molecules that make up the bag magically seal back around the pencil and prevent water from leaking out. Now, that's the "Spear-It" of science! Sound impossible? Try it—over the sink the first time and then over your friend's head . . . just for fun.

5. This is just like a Las Vegas–style magic act where the magician jabs swords through the cabinet, yet the beautiful assistant comes out unscathed. Okay, maybe not, but you have to agree that it's amazing that the water stays in the bag. Be careful not to push the pencils all the way through the holes or your science magic trick will quickly turn into a "clean-it-up" activity.

6. When you are finished, hold the bag over the sink or a bucket and remove the pencils. Toss the bag in the recycling bin and dry the pencils.

TAKE IT FURTHER

Try experimenting with plastic bags of different sizes and thicknesses. The thicker the bag, the harder it is to get the pencil to pass through. For a really thin bag, use a plastic bag from the produce section of the grocery store.

Experiment with different sizes and shapes of pencils. Some pencils have flat edges while others have perfectly round, smooth edges. Which type of pencil works best?

This next gem is an extension of the "Baby Diaper Secret" activity. Collect the superabsorbent powder from the lining of two or three diapers. After all of the pencils are pushed through the bag, carefully open the bag and sprinkle in the superabsorbent powder. Give the powder a few seconds to solidify the water and remove the pencils.

Use "The Leakproof Bag" as an object lesson for a message on school spirit and leadership.

Let the bag of water represent the student body and use the pencils to demonstrate the excitement of school spirit (spear-it!).

Each pencil represents a different element of school spirit—teamwork, pride, unity, attitude, dedication, and fun.

As the pencils pass through the water, the student body helps to magnify the message of involvement and participation in school activities throughout the year. What happens when the spirit pencils are removed? Is it possible to stop the leaks? How do we keep the enthusiasm alive?

Collect the superabsorbent powder from the "Baby Diaper Secret" in a cup and label it "LEADERSHIP." With just the proper amount of leadership, the student body is forever changed.

WHAT'S GOING ON HERE?

The zipper-lock plastic bag you used was most likely made out of a **polymer** called low-density polyethylene (LDPE). It's one of the most widely used packaging materials in the world. LDPE is low in cost, lightweight, durable, a barrier to moisture, and very flexible.

Think of the polyethylene molecules as long strands of freshly cooked spaghetti. The tip of the sharpened pencil can easily slip between and push apart the flexible strands of spaghetti, but the strands' flexible property helps to form a temporary seal against the edge of the pencil. When the pencil is removed, the hole in the plastic bag remains because the polyethylene molecules were pushed aside permanently and the water leaks out.

As you might have discovered, it's much easier for the stretched plastic to seal around the smooth sides of a round pencil than the straight edges found on other pencils. Hopefully you discovered this tip during practice and not while the bag was precariously positioned over someone's head.

THE **EGG DROP**

The Egg Drop is one of my all-time favorite science demonstrations. It's a combination of strategy and skill . . . and just a little luck. The goal is to get an egg to drop into a glass of water. Sound easy enough? Did I mention that the egg is perched high above the water on a cardboard tube and that a pie pan sits between the tube and the water? Still think it's easy? Sir Isaac Newton does. Once you try It, you'll be hooked!

WHAT YOU NEED

Large eggs
(buy a dozen because
you need the practice)

Cardboard tube from an
empty roll of toilet paper

Metal pie pan

Pitcher of water

Large drinking glass

Oh, you might need a few
paper towels to clean up
your practice mess!

LET'S TRY IT!

1. Pick a sturdy table or counter surface to perform the demonstration. Fill the drinking glass about three-quarters full with water and center the pie pan on top of the glass. Place the cardboard tube vertically on the pie pan, positioning it directly over the water. Carefully set the egg on top of the cardboard tube.

2. Explain to your audience that the goal is to get the egg into the glass of water, but you're not allowed to touch the egg, the cardboard tube, or the glass of water. The only thing left for you to touch is the pie pan. What would you need to do to move the pie pan and cardboard tube out of the way in order for the egg to fall into the glass of water? That's right . . . you're going to invoke Sir Isaac Newton's First Law of Motion and smack the pie pan out of the way. Don't do it just yet . . . read the next step.

3. Stand directly behind the Egg Drop setup. If you're right handed, hold your right hand straight out like you were going to karate chop something. Position

your hand about 6 inches away from the edge of the pan. The idea is to hit the edge of the pie pan with enough force to knock the cardboard tube out from under the egg. Gravity will do the rest as the egg falls directly into the glass of water.

4. Shoot both hands up high over your head in celebration of your latest science miracle.

TAKE IT FURTHER

Try testing longer cardboard tubes from a roll of paper towel, different size glasses, or different size eggs. Do small eggs work as well as jumbo eggs?

The true Egg Drop connoisseur will never be content with a single egg falling into a single glass. The temptation is just too great to push the envelope and find a way to position two eggs, side by side, and attempt a drop. When it works (and it will), you'll discover that two eggs just aren't enough. After searching day and night for weeks on end (or maybe you'll just find one lying around the house), you'll find the perfect tray to hold three cardboard tubes and three eggs. It's no longer a science experiment . . . it's an obsession with the law of inertia and gravity. Wake the kids, phone the neighbors—this is going to be something special.

WHAT'S GOING ON HERE?

Credit for this one has to go to Sir Isaac Newton and his **First Law of Motion**. Newton said that objects in motion want to keep moving and objects that are stationary want to stay still—unless an outside force acts on them. So, since the egg is not moving while it sits on top of the tube, that's what it wants to do—not move. You applied enough force to the pie pan to cause it to zip out from under the cardboard tube (there's not much friction between the surface of the pan and the water container). The edge of the pie pan hooked the bottom of the tube, which then sailed off with the pan. Basically, you knocked the support out from under the egg. For a brief nanosecond or so, the egg didn't move because it was already stationary (not moving). But then, as usual, the force of gravity took over and pulled the egg straight down toward the center of the Earth.

Also, according to Mr. Newton's First Law, once the egg began moving, it didn't want to stop. The container of water interrupted the egg's fall, providing a safe place for the egg to stop moving so you could recover it unbroken. The force of gravity on the egg caused the water to splash out, and the audience burst into spontaneous applause.

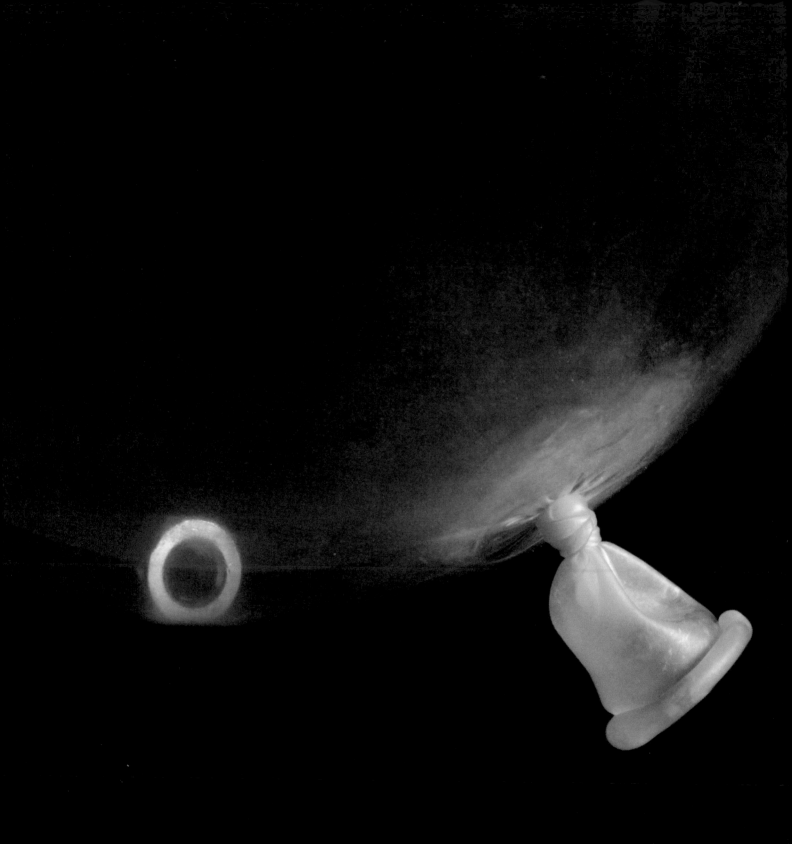

SCREAMING BALLOON

Here's an easy-to-do experiment using only a balloon and a hex nut from the hardware store. Be sure to buy enough supplies for all of your friends because there's nothing better than a room filled with screaming balloons. Oh, and it's also a great way to learn about the science of sound.

LET'S TRY IT!

1. Squeeze the hex nut through the mouth of the balloon. Make sure that the hex nut goes all the way into the balloon so that there is no danger of it being sucked out while blowing up the balloon.

2. Blow up the balloon, but be careful not to overinflate the balloon, as it will easily burst. Tie the end of the balloon and you're ready to go.

3. Grip the balloon at the stem end as you would a bowling ball. The neck of the balloon will be in your palm and your fingers and thumb will extend down the sides of the balloon.

4. While holding the balloon palm down, swirl it in a circular motion. The hex nut may bounce around at first, but it will soon begin to roll around the inside

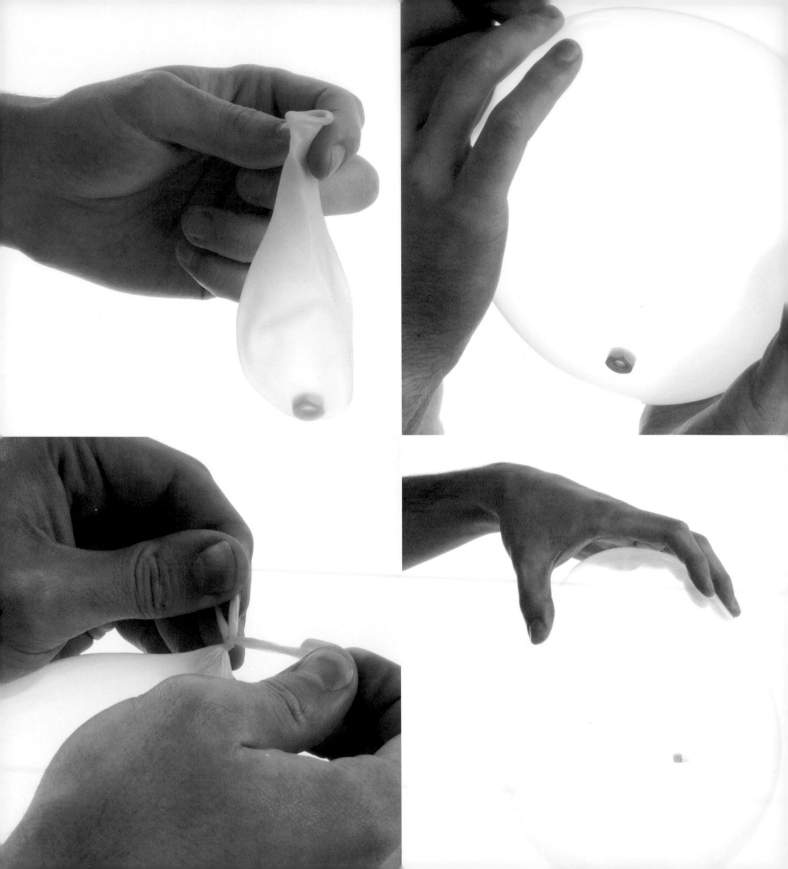

of the balloon. What is that sound? Could the balloon be screaming? The sound every parent loves . . .

5. Once the hex nut begins to spin, use your other hand to stabilize the balloon. Your hex nut should continue to spin for 10 seconds or more.

TAKE IT FURTHER

What happens when you change the size of the balloon or the size of the hex nut? Try using a marble in place of a hex nut. Does the marble make the balloon "scream?" (Hint: Librarians like the marble version of this activity.) Experiment with other objects whose edges may vibrate against the balloon.

Take a trip to the moon in order to try the experiment in an environment with less gravity. How does this affect the outcome of the experiment? Does the hex nut spin for a longer period of time? Come back to Earth and hold a press conference.

WHAT'S GOING ON HERE?

This is actually a 2-for-1 experiment—you're learning about the science of motion and sound. To understand how and why it works, you have to look at the forces that are acting on the hex nut. The shape of the balloon makes the hex nut move in a circular path. Otherwise, the hex nut would want to continue to move in a straight line. Another force to consider is friction. There's very little friction between the edge of the hex nut and the balloon. More friction would cause the hex nut to slow down and stop.

The real force in action here is a centripetal, or center-seeking, force. **Centripetal force** is the inward force on a body that causes it to move in a circular path. The old concept of "centrifugal force," an outward or center-fleeing force, has been largely replaced by a more modernistic understanding of centripetal force.

A hex nut has six sides, and these flat edges cause the hex nut to bounce or vibrate inside the balloon. The screaming sound is made by the sides of the hex nut vibrating against the inside wall of the balloon.

BURNING MONEY

Do you have money burning a hole in your pocket? It's probably not a wise idea to soak a $20 bill in a flammable liquid and set it on fire, but that's what you'll have to do in this science demonstration. Sure, you could just use a $1 bill, but then you won't sweat as much.

WHAT YOU NEED

70% rubbing alcohol
(read the labels in the store to determine the alcohol content)

Water

Tongs

Lighter or match

Safety glasses

Fire extinguisher

WARNING! IMPORTANT SAFETY RULES

This demonstration should not be attempted without strict adult supervision.

LET'S TRY IT!

1. Start by preparing a water-alcohol mixture by combining 3 ounces (90 mL) of 70% rubbing alcohol with 1 ounce (30 mL) of water. Make sure to stir the mixture thoroughly.

2. Rule #1: Never use your own money. Borrow a $20 bill from your friend. If you can get away with this, you have incredible skills of persuasion. Otherwise, cough up your own money.

3. Dip the bill into the mixture of water and rubbing alcohol using the tongs, and make sure the bill gets completely soaked.

4. Remove the bill using the tongs and gently shake off any excess liquid.

5. Move the water-alcohol mixture to a safe place (away from the area where you are going to light the bill on fire).

6. Hold one end of the bill with tongs and light the bottom of the bill with a lighter. The bill will *look* like it's burning, but it shouldn't burn (famous last words). When the flame is completely extinguished, it's safe to touch the money. You'll find that the money is even cool to the touch.

Alcohol burns with an almost invisible blue flame. One trick is to add a little table salt to the water-alcohol mixture to give the flame a more yellowish color and make it more visible.

You can also try to change the ratio of rubbing alcohol to water to see how it affects the way the bill burns, but you're likely to accidentally burn up your dollar bill.

WHAT'S GOING ON HERE?

By now you've probably guessed that the money *will* actually burn if you dip it into a pure alcohol solution. The secret, of course, is the addition of water to the mixture. The water from the water-alcohol mixture evaporates and absorbs much of the heat energy that is generated when you ignite the bill. The water is first heated to its boiling point and then vaporized by the heat of combustion from the burning alcohol. The evaporation of the water keeps the temperature below the ignition temperature of paper, which is 233°C or 451°F. If you read *Fahrenheit 451*, a novel by Ray Bradbury about book burning, you will recognize this famous temperature. If you reduce the amount of water in the mixture, the paper money is likely to char or even catch on fire.

MENTOS **GEYSER** EXPERIMENT

It's been called the "vinegar and baking soda" reaction for a new generation. While science teachers have been dropping candies and mints into 2-liter bottles of soda for years in an effort to release all of the dissolved carbon dioxide, the Mentos and Diet Coke reaction became world famous in 2005. Fueled by hundreds of blogs and popular online sharing sites like YouTube, this once obscure reaction became an Internet sensation, and the enthusiasm for dropping Mentos into soda continues to grow. Once you get past the initial gee-whiz factor, there's some amazing science behind a carbonated beverage and a chewy mint.

WHAT YOU NEED

A roll or box of Mentos chewy mints (stick with the standard mint flavor for now)

2-liter bottle of diet soda (either diet or regular soda will work for this experiment, but diet soda is not sticky when you're cleaning it up, and it will usually create a bigger blast)

Sheet of paper to roll into a tube

Steve Spangler's Geyser Tube toy (optional . . . but highly recommended!)

LET'S TRY IT!

1. This activity is probably best done outside in the middle of an abandoned field or on a huge lawn.

2. Carefully open the bottle of diet soda. Again, the choice of diet over regular soda is purely a preference based on the fact that erupting regular soda becomes a sticky mess to clean up because it contains sugar. Diet soda uses artificial sweeteners instead of sugar, and consequently, it's not sticky. Later on in the experiment, you'll be invited to compare the geyser power of diet versus regular soda, but for now we'll start with a 2-liter bottle of diet soda.

3. Position the bottle on the ground so that it will not tip over.

4. Let's start with seven Mentos for our first attempt. The goal is to drop all seven Mentos into the bottle of soda at the same time (which is trickier than you might think). One method for doing this is to roll a piece of paper into a tube just big enough to hold the loose Mentos. Other methods include using a large plastic test tube to hold the Mentos or using my Geyser Tube toy invention,

which was created to solve this very problem. Assuming that you're using the paper tube method, you'll want to load the seven Mentos into the tube, cover the bottom of the tube with your finger, and position the tube directly over the mouth of the bottle. When you pull your finger out of the way, all seven Mentos should fall into the bottle at the same time.

5. Enough waiting . . . this anticipation is killing me. 3-2-1 drop the Mentos!

6. This final step is very important . . . run away! But don't forget to look back at the amazing eruption of soda.

7. If spectators were watching your exploits, someone is bound to yell out, "Do it again!" and that's exactly what you're going to do.

Simply dropping Mentos into a bottle of soda to make a geyser isn't really science—it's just a fun trick to do in the backyard. The real learning takes place when you start to change one variable at a time to see how it affects the performance of the geyser.

Check out the "Science Fair Connection" on page 149 for great ideas on how to measure the height of the geyser and how to determine the best ingredients to use to make the highest-shooting soda geyser.

WHAT'S GOING ON HERE?

Why do Mentos turn ordinary bottles of diet soda into geysers of fun? The answer is a little more complicated than you might think. Let's start with the soda . . .

Soda pop is made of sugar or artificial sweetener, flavoring, water, and preservatives. The thing that makes soda bubbly is invisible carbon dioxide (CO_2), which is pumped into bottles at the bottling factory using lots of pressure. If you shake a bottle or can of soda, some of the gas comes out of the solution and the bubbles cling to the inside walls of the container (thanks to tiny pits and imperfections on the inside surface of the bottle called **nucleation sites**). When you open the container, the bubbles quickly rise to the top pushing the liquid out of the way. In other words, the liquid sprays everywhere.

Is there another way for the CO_2 to escape? Try this. Drop an object like a raisin or a piece of uncooked pasta into a glass of soda and notice how bubbles immediately form on the surface of the object. These are CO_2 bubbles leaving the soda and attaching themselves to the object. For example, adding salt to soda causes it to foam up because thousands of little bubbles form on the surface of each grain of salt. This bubbling process is called **nucleation**, and the places where the bubbles form, whether on the sides of the can, on an object, or around a tiny grain of salt, are the nucleation sites.

Why are Mentos so special?

The reason why Mentos work so well is twofold—tiny pits on the surface of the mint, and the weight of the Mentos itself. Each Mentos mint has thousands of tiny pits all over the surface. These tiny pits act as nucleation sites—perfect places for CO_2 bubbles to form. As soon as the Mentos hit the soda, bubbles form all over the surfaces of the candies and then quickly rise to the surface of the liquid. Couple this with the fact that the Mentos candies are heavy and sink to the bottom of the bottle and you've got a double whammy. The gas released by the Mentos literally pushes all of the liquid up and out of the bottle in an incredible soda blast.

Science Fair Connection

You might be asking yourself, "Can I use the Mentos Geyser for my science fair project?" The answer is YES, but you'll need to learn how to turn a cool science activity into a real science experiment. The secret is to turn your attention away from the spraying soda and concentrate on setting up an experiment where you isolate a single variable and observe the results.

To get the best results in a science experiment you need to standardize the test conditions as much as possible. The biggest challenge in the Mentos Geyser experiment is finding a consistent way to drop the Mentos into the soda every time. The original reason I invented the Geyser Tube toy was to find a way to standardize the actual drop of the Mentos. If you're not using the Geyser Tube, make sure to come up with your own method for dropping the Mentos into the soda the same way each time.

Measuring the Height of the Geyser

To make any of these tests meaningful, you need to find a way to measure the height of the eruption. A friend or parent with a video camera is a great way to watch and document the results of your experiment, but you'll also need some specific measurements or data. Try placing the soda bottle next to the wall of a brick building (after getting permission from the building's owner). Measure the height of the geyser by counting the number of bricks that are wet once the geyser stops. If you want a more specific measurement, use chalk to mark off 1-foot increments on the brick

wall before you drop the Mentos into the bottle of soda. Make comparisons, create a chart with your data, and draw some conclusions. Be sure to thank the building's owner and to hose off the wall of the building when you are finished!

Measuring the Volume of the Geyser

If you want to examine the volume of the geyser instead of the height, make note of the volume of a full bottle of soda before you drop the Mentos into it. (Okay, it's a trick question because a 2-liter bottle of soda holds . . . 2 liters!) Once the geyser stops, pour out the remaining contents of the bottle and measure how much liquid is left. You could use a beaker or a graduated cylinder to measure the remaining liquid in milliliters. Remember that 1 liter is equivalent to 1000 mL. Subtract the remaining amount of liquid from the original volume of the bottle to calculate the volume of the geyser. Then make comparisons, create a chart with your data, and draw some conclusions.

How Many Mentos Work Best?

This has to be the number one question everyone asks about this experiment. What is the best number of Mentos to use to make the highest-shooting geyser? This is a great topic for a science project—you'll need lots of soda and Mentos, and a few friends to help record all of the data.

Be sure that the soda bottles are all the same brand and type. It's also important that all of the test bottles are stored in the same place so that the liquid in each bottle is the same temperature.

Line up a row of ten 2-liter bottles against a brick wall (see "Measuring the Height of the Geyser"). Each bottle will receive a different number of Mentos. Drop one Mentos into the first bottle and record the height by counting the wet bricks (or set up your own scale behind each soda bottle). Drop two Mentos into the second bottle, and so on until you've completed all ten bottles.

Of course, this could go on forever, but you'll start to see a trend in your data that shows the maximum height of the geyser for a certain number of Mentos. Many soda geyser-ologists believe that seven Mentos produce the highest-shooting geyser. Using any more than seven Mentos is just a waste, according to these soda-soaked science enthusiasts. What do your results reveal about the effect of the number of Mentos on the height of the geyser?

The Brand Test

You guessed it . . . it's time to put your favorite soda to the test. Does one brand produce higher-flying geysers? How does generic soda stack up against the big name brands? If you're doing a science fair project, your initial question might be, "What is the effect of the brand of soda on the height of the geyser?"

Use your data from the previous test to determine the standard number of Mentos to use for this test. The only variable you'll change in this test is the brand of soda while everything else remains the same (the number of Mentos and the amount of soda). Again, make sure all of the soda is at the same temperature because temperature plays an important role in the reaction. The brand of soda is the only thing that changes (the variable).

Just think . . . your results could help determine the next Mentos Geyser craze!

The Temperature Test

What is the effect of temperature on the height of the geyser? Does warm soda shoot up higher than cold soda? The key is to keep every launch fair and to make sure the only variable is the temperature of the soda. You'll need a thermometer to record the temperature of the soda just before you launch it.

To enforce the fairness factor, you must stick with one brand of soda for the entire test. Let's use Diet Coke in this example. You'll want to purchase three bottles of Diet Coke and two rolls of Mentos. You're going to set up three tests—warm soda, room temperature soda, and cold soda. Place one bottle of Diet Coke in the refrigerator and let it sit overnight. Place the second bottle in a place where it can reach room temperature overnight. There are two safe ways to warm the other bottle of soda. The simplest method is to let the unopened bottle sit in the sun for several hours. You can also place the bottle of unopened soda in a bucket of warm water. Never use a stove or microwave to heat a bottle of soda.

It's time to return to your launching site. Check to make sure your measuring scale is in place (counting bricks or using an alternative scale against the wall). Let's start with the bottle of cold Diet Coke. Open the bottle and dip the thermometer down into the soda. Record the temperature. Load seven Mentos into your paper roll and drop them into the soda. Immediately record the data for the cold soda test. Repeat the same procedure for the bottle of soda at room temperature and for the bottle of warm soda. It's important to use the same number of Mentos for each test and to drop them the same way.

No matter which brand of soda you tested, the warm bottle probably produced the highest-shooting geyser. Warm soda tends to fizz much more than cold soda. Why? The answer lies in the solubility of gases in liquids. The warmer the liquid, the less gas can be dissolved in that liquid. The colder the liquid, the more gas can be dissolved in that liquid. This is because as the liquid is heated, the gas within that liquid is also heated, causing the gas molecules to move faster and faster. As the molecules move faster, they diffuse out of the liquid, leaving less gas dissolved in that liquid. In colder liquids the gas molecules move very slowly, causing them to diffuse out of the solution much more slowly. More gas tends to stay in solution when the liquid is cold. This is why at the bottling plant CO_2 is pumped into the cans or bottles when the fluid is just above freezing—around 35° F. This low temperature allows the maximum amount of CO_2 to dissolve in the soda, keeping the carbonation levels as high as possible.

The Big Blast

After completing all of these tests, you've become somewhat of a Mentos Geyser expert who has the research to support the answer to the question, "How can you make the highest-shooting Mentos geyser?" Each test isolated an independent variable, and combining all of the information you discovered into one launch is a great way to wrap up your science fair project. For example, based on your individual test results, you might have arrived at this recipe for the best Mentos Geyser:

Use a bottle of Diet Coke

Make sure the soda is at least 85°F

Drop seven Mentos into the soda all at the same time

By using the scientific method and some critical thinking skills, you've successfully turned a great gee-whiz science trick into a research-based science fair project.

Mentos Geyser History—From Obscurity to Instant Celebrity

As strange as it might sound, the Mentos Geyser never actually started out using Mentos chewy mints. This science demonstration was popular among chemistry teachers back in the 1980s using a roll of Wintergreen LifeSavers and a pipe cleaner. Teachers threaded the roll of Wintergreen LifeSavers onto a pipe cleaner as an easy way to drop all of the LifeSavers into the soda at the same time. Within seconds of dropping the candies into the soda, a huge geyser would erupt from the bottle.

However, by the end of the 1990s, the manufacturer of Wintergreen LifeSavers increased the size of the mints (no one was ever certain why this happened), making the diameter of the candy too large to fit into the mouth of the soda bottle. Science teachers started

experimenting (as they like to do) with other candies and mints that would have the same effect when dropped into a bottle of soda. As luck would have it, the solution to the problem was within arm's reach of the Wintergreen LifeSavers in the candy aisle—it was Mentos chewy mints.

Because Mentos mints didn't have holes in the middle like LifeSavers, getting them into the bottle was tricky. Everyone found their own method of quickly dropping the Mentos into the soda. Some people fashioned a tube out of paper while others used a piece of plastic tubing to load the Mentos. At the time, my solution was to load the Mentos candies into something called a Baby Soda Bottle—a test tube–like container that held an entire roll of Mentos perfectly. Oddly enough, this container was actually a "pre-form" or 2-liter soda bottle before it was blown up into a big bottle. That's why it's called a Baby Soda Bottle.

However, I must admit that even with the Baby Soda Bottle method, the results were not very consistent and it was challenging to get away from the bottle before it exploded. So, I solicited help from our creative team at Steve Spangler Science to come up with a Geyser Tube—a better, more consistent way to drop the Mentos into the bottle. Better yet, if we could trigger the drop of the Mentos from a distance, we wouldn't get as wet.

The next few months were spent building trigger devices ranging from plastic tubes with sliding doors to magnets that held metal stoppers in place to an elaborate battery-operated switch that was triggered by a motion detector. We even played with ways of using the Geyser Tube to trigger multiple soda geysers in a method similar to a Rube Goldberg machine. But the bottom line was that we needed to find a way to standardize the drop of the Mentos.

As they say, the simplest design usually turns out to be the best and most elegant solution to the problem. The winning Geyser Tube design was a clear plastic tube with a special fitting that twisted onto any soda bottle.

The trigger pin at the bottom of the tube prevented the Mentos from falling into the bottle until you pulled the string attached to the pin. The moment the pin was pulled, a slider ring resting above the pin fell into place and covered the holes where the trigger pin once was, and the Mentos dropped into the soda. But there was one added bonus . . . the restricted hole at the top of the plastic tube helped to build up more pressure in the bottle and launched the soda 30 feet into the air.

Fortunately, the maker of Mentos (Perfetti Van Melle) also liked the design, and we launched the Mentos Geyser Tube toy at the New York Toy Fair in February 2007. The Geyser Tube toy is currently available in toy stores and mass-market retailers throughout the country thanks to our distributor, Be Amazing Toys!

The Mentos Geyser became one of my featured demonstrations both on television and during my live stage presentations. While I had performed variations of the Mentos Geyser experiment on television many times from 2001 to 2004, my performance of the demo in September of 2005 in the backyard of NBC affiliate KUSA-TV in Denver proved to be the tipping point as the demo went from relative obscurity to Internet sensation.

My cohost for the KUSA-TV science segment was the lovely Kim Christiansen. During the commercial break, I told Kim what was going to happen and reminded her to

pull her hand out of the way of the erupting geyser and to run backward. Unfortunately, Kim got so caught up in the fun that she forgot to do both . . . and got soaked in Diet Coke on live television. To add insult to injury, she did it two more times, each time getting covered in more soda, until her once pink dress was more Coke-colored than pink.

KUSA-TV News posted that original video on their website along with my blog post titled, "News Anchor Gets Soaked!" Within a few weeks, links to the video and my blog entry numbered in the thousands. I also posted the video on a new online video sharing site called You-Tube (YouTube was only 7 months old at the time), and as they say, the rest is history. Within the next 12 months, over 800 Mentos Geyser-related videos were posted on YouTube, making the demo one of the most popular pop-culture science experiments in recent history.

You know the Mentos Geyser is a popular experiment when a producer from ABC's *Who Wants to Be a Millionaire* calls for help writing a question. Here's the question we came up with:

In an experiment popularized online, what candy creates an explosive geyser when dropped into a 2-liter Diet Coke bottle?

A) Skittles
B) Mint Mentos
C) Atomic Fireballs
D) Lemon Heads

The question was asked on a special College Week episode of *Who Wants to Be a Millionaire*. The participant got it right for $8,000, saying: "I saw it on TV and I bought Mentos and a 2-liter bottle of Diet Coke . . . so I'm going to go with Mentos. That's my final answer." The contestant ended up doing really well, going all the way to the $250,000 question, but he walked away with $125,000.

THANKS FOR CREATING A FEW
UNFORGETTABLE
LEARNING EXPERIENCES

I can remember the first time one of my elementary teachers demonstrated what I considered to be a real science experiment in front of the class. Unfortunately, I had to wait until I was in fifth grade, but it was worth the wait. I remember vividly the day that Mr. London took off his suit jacket (yes, he wore a suit), put on an apron, and unearthed a model of a volcano from the storage closet in the back of the room. This was completely out of the ordinary since all of our science instruction up to this point had come from an outdated textbook.

"Shhhh! He's got a volcano!" whispered one of my friends. I heard someone ask, "What do you think he's going to do?" The anticipation was almost too great.

All eyes were fixed on his every move as Mr. London placed the volcano on the table in the front of the room. He returned to the closet to get more supplies, and I remember thinking, "He's going to get some vinegar and baking soda." While I had never tried the reaction myself, I had read about making my own volcano in a *Mr. Wizard Kitchen Chemistry* book.

But he didn't grab any vinegar or baking soda. Instead, Mr. London pulled out a bottle of some purple powder and a small vial that contained a clear liquid. He poured the purple chemical into the opening of the volcano and then added a few drops of the clear, syrupy liquid. Within a few seconds, the volcano started to sputter and smoke. The thick white smoke quickly turned into sparks and little tiny balls of fire. The volcano was really erupting . . . and then the whole thing caught on fire! I remember thinking *this is the greatest day of my life!*

Everything switched into high gear from this point forward as Mr. London ran to get his suit jacket in order to cover the fire-breathing volcano. Smoke bellowed from his dark blue jacket. While none of us dared to cheer out loud, we all wanted to explode with excitement. The fire was out in a matter of seconds, but the smoke forced everyone to move quickly to the hallway. When we returned to the classroom, the once nicely painted volcano was completely black and there was a deep, circular burn mark in the middle of the table.

The burn mark on the table was more than a blemish, it was a reminder to everyone that science is filled with wonder and surprises. Mr. London probably expected lots of *ooohs* and *ahhhs* that day, but what he ended up with was something far more lasting—an unforgettable experience. As cliché as it might seem, that experience changed my view of science forever.

I returned to my elementary school in Denver, Colorado, in 1998, nearly 20 years after my fifth grade flaming volcano experience. I was invited to speak at the school assembly as part of a kickoff for their science fair. Mr. London had long since retired, and none of the current staff knew anything about the famous volcano experiment of 1978. But, after hearing the story, a young lady who worked as the media specialist took me up to the library on the second floor. Tucked away in the corner was an old table that was covered with magazines and books. She pushed away a stack of books in the very middle to reveal a large, circular burn mark. Unforgettable.

My hope for you (the person who stepped in every time the experiment called for adult supervision) is that you'll pause for just a minute to think back on all of the experiences that you created for your young scientist by merely doing some of the experiments in this book. Some of the experiments that you tried from the preceding pages worked beautifully, while others might not have gone exactly as planned. Regardless of the outcome, the true test of your success was evident in the smiles and expressions of wonder on the faces of your children, grandchildren, students, or just friends who came over to see what you had up your sleeve. Through your experimentation, you created a wealth of unforgettable learning experiences . . . and no one could have asked for more.

Keep sharing the wonder of learning.

—Steve